Advanced Sciences and Technologies for Security Applications

Indexed by SCOPUS

The series Advanced Sciences and Technologies for Security Applications comprises interdisciplinary research covering the theory, foundations and domain-specific topics pertaining to security. Publications within the series are peer-reviewed monographs and edited works in the areas of:

- biological and chemical threat recognition and detection (e.g., biosensors, aerosols, forensics)
- crisis and disaster management
- terrorism
- cyber security and secure information systems (e.g., encryption, optical and photonic systems)
- traditional and non-traditional security
- energy, food and resource security
- economic security and securitization (including associated infrastructures)
- transnational crime
- human security and health security
- social, political and psychological aspects of security
- recognition and identification (e.g., optical imaging, biometrics, authentication and verification)
- smart surveillance systems
- applications of theoretical frameworks and methodologies (e.g., grounded theory, complexity, network sciences, modelling and simulation)

Together, the high-quality contributions to this series provide a cross-disciplinary overview of forefront research endeavours aiming to make the world a safer place.

The editors encourage prospective authors to correspond with them in advance of submitting a manuscript. Submission of manuscripts should be made to the Editor-in-Chief or one of the Editors.

More information about this series at https://link.springer.com/bookseries/5540

Juan Cayón Peña
Editor

Security and Defence: Ethical and Legal Challenges in the Face of Current Conflicts

Springer

Editor
Juan Cayón Peña
Campus de la Berzosa
Universidad Antonio de Nebrija
Hoyo de Manzanares, Spain

ISSN 1613-5113 ISSN 2363-9466 (electronic)
Advanced Sciences and Technologies for Security Applications
ISBN 978-3-030-95938-8 ISBN 978-3-030-95939-5 (eBook)
https://doi.org/10.1007/978-3-030-95939-5

This Springer imprint is published by the registered company Springer Nature Switzerland AG
The registered company address is: Gewerbestrasse 11, 6330 Cham, Switzerland

Foreword: Rule of Law. Looking for a Warrior Ethos

Let me start this by thanking the people that had the idea of this Foreword, Prof. Juan Cayon and Prof. Jesus Martin-Ramirez from Nebrija University. Thanks to both of you for this opportunity given to an old soldier. Professor Cayón, an old friend of mine, suggested that I could use my own experience to address, thru the screen with my personal view on the importance of military people following the rule of law and, at the same time, trying to keep an ethical behavior. From my perspective is an important duty, particularly in these times we are forced to confront with the threats of liquid postmodernism. My reply, there was no other possible, was an airborne "HOAH!", or much polite and much better "AYE AYE, Sir, I will do my best".

Armed Forces in Western democracies are established and organized in such a way that not being democratic on their own, in their daily work (nobody asks a soldier if he or she is happy with a certain task), their chain of command respects and follows the rule of law. This term, rule of law, has been described in varying terms, but I believe that a comprehensive definition, such as the one used by the United Nations, is useful in this context. They define rule of law as a principle of governance in which all persons, institutions, and entities, public and private, including the different nations involved, are accountable to laws that are publicly promulgated, equally enforced and independently adjudicated, and which are consistent with international human rights norms and standards. It requires, as well, measures to ensure adherence to the principles of supremacy of law, equality before the law, accountability to the law, fairness in the application of the law, separation of powers, participation in decision-making, legal certainty, avoidance of arbitrariness, and procedural and legal transparency.

All this is very important because the only way the Armed Forces can be sure they use the power given to them in the best interest of the people is by subordinating their possibility to exercise military power to the three powers of state (legislative, executive, and judiciary). Please be aware that the actual subject of the national sovereignty in most of the Western nations is, precisely, "we, the People".

Back in the seventeenth century when sovereignty resided upon the King, we use to refer to the Armed Forces as "His Majesty's Armies". Today we keep the traditional name, but militaries are no longer subject to the King or the Queen. Today in most of these nations, including my own, the three powers of state are representing national sovereignty in the name of the people and so it is the government elected by them the one that sets the terms of the national defense and analyzes and faces the threats and risks to the subsistence of Spain as an advanced democratic society. In Spain, we used to say and sometimes we convince our political masters that policy is for them as much as strategy is for military people.

This is the only way I see to guarantee that the monopolistic use of force granted to the Armed Forces can provide institutional stability and ensure the existence of a nation-state. To achieve such a task, Armed Forces need to arrive at a kind of legal contract between themselves and the nation. This contract or clause is what we call "subordination of military affairs to political power".

Many nations use secular constitutions, statutes, and mechanisms (law enforcement, courts, and institutions to correct individual criminal behavior) to implement rule of law. But other nations use differing implementation methods.

In the Middle East, for example, the structure of the legal system is derived from a combination of systems, including religion and tribal practice, to form formal and informal legal mechanisms. In states with a strong Islamic influence, personal issues, such as divorce and marriage, are resolved in sharia courts. The judges in criminal courts may be educated to approach criminal matters differently than judges trained in sharia law. In Central and Southern America, what Spaniards refer to as Hispanic-Americas, citizens believe that they do not have a voice or the ability to obtain justice from a system permeated by corruption, judicial failures, repressive police tactics, and the legal marginalization of the majority of the population. Citizens in Central Asia, most notably the Caucasus states, assume that they are governed by institutions that are inept, corrupt, and rife with nepotism.

In some parts of the world, rule of law appears under the guise of a strong authoritarian ruler exercising great influence over the "independence" of the judiciary system. Often rulers with a strong "law and order mentality" impede social change that may threaten their holds on power.

In the Spanish case, when referring to the government and its executive powers, our constitution says in its Article 97th, "the Government directs internal and foreign policy, the civil and military administration and the defense of the state…". Spanish constitution tasks Armed Forces, constituted by the Army, the Navy, and the Air Force to fulfill their mission: to guarantee sovereignty and independence of Spain, to defend its territorial integrity and constitutional order. Constitution also instructs them to do fulfill that mission in accordance with the law and the principles of the constitution itself. This is to say that Armed Forces must subordinate their actions to the political powers and to adjust them to legality in order to legitimize their possibility of using lethal force while marrying up Armed Forces with the society they serve to.

In any case, there are two/three different sets of rules the Armed Forces must follow.

On the one hand, they must follow their national law, constitution, and the different laws derived from her. The capacity to use arms to oppose exterior and inner threats to the nation in this kind of monopolistic exercise makes so important for militaries to follow the rules established by the government and the law.

In both cases, the rule of law is hard to transform into tangible successes on the ground, but it is still very important to develop, particularly when we are trying to set the ground for a "comprehensive civil-military approach" as we, members of NATO, usually do. This is the reason behind military intent to prepare our ranks to confront this challenge when we deploy to places like Afghanistan in which stability and rule of law are critical to security in the region and throughout the globe. The need to strengthen the rule of law will exist wherever people and their governments cope with conflict and instability. No wonder why President Barack Obama stated when addressing the transition from International Security Assistance Force (ISAF) to the Resolute Support Mission: "Western commitment to the rule of law is fundamental to our efforts to build an international order that is capable of confronting the emerging challenges of the 21st century". As it happened in Afghanistan, Western Nations will continue to engage in states at risk of failure, in failed states where the central government is so weakened that the people have virtually returned to the natural state, in states emerging from long periods of conflict such as Afghanistan, and in states in peaceful postconflict rebuilding periods.

On the other hand when in operations abroad, they need to cope with the law and constitution of the countries in which they are deployed. Armed Forces must follow the Law of Conflict also known as the LAW OF ARMED CONFLICT or the "Geneva Conventions" that comprise four treaties and three additional protocols. All of them together are able to establish the standards of international law for humanitarian treatment in wartime.

It was a Swiss businessman, Mr. Henry Dunant, Co-recipient of the Nobel Peace Prize in 1901 when visiting wounded soldiers after the Battle of Solferino in 1859 that realized the lack of facilities, personnel, and medical aid available to soldiers. His wartime experience inspired him to propose a permanent relief agency for humanitarian aid in times of war and a treaty recognizing the neutrality of this agency and allowing her to provide aid in a combat zone. This proposal led to the establishment of the Red Cross and later on to the Geneva Convention as the first codified international treaty that covered the sick and wounded soldiers on the battlefield. Years later, the Swiss government invited European countries, as well as the USA, Brazil, and Mexico, to attend an official diplomatic conference. Sixteen countries replied, and the conference adopted the first Geneva Convention signed by twelve countries Switzerland, Belgium, Denmark, France, Italy, the Netherlands, Portugal, and Spain, among them.

In reality, the term Geneva Convention usually denotes the agreements of 1949, negotiated in the aftermath of the Second World War which updated the terms of the previous treaties and added two new conventions. The Geneva Conventions extensively defined the basic rights of wartime prisoners (civilians and military personnel), established protections for the wounded and sick, and established protections for the civilians in and around the area of operations. The treaties of 1949 were ratified, in their entirety or with reservations, by 196 countries. Moreover, the Geneva Conventions also define the rights and protections afforded to non-combatants but they do not address the use of weapons of war, which are the subject of the Hague Conventions, nor the biochemical warfare that belongs to the Geneva Protocol.

For militaries across the world, it is very important to comply with the Geneva Conventions. It is important for the soldier, for the commander and even for the country because every member of the Armed Forces, whatever his or her rank is, has a personal responsibility not only to comply with the law of armed conflict but to ensure that it is complied with by others and to take action in the event of violations. There is no possibility for any member of the Armed Forces to invoke superior orders in their defense. Violations of the law are considered war crimes and will be prosecuted in national courts or in international tribunals such as the International Criminal Court.

Despite the difficulties of formulating threats and risks in an environment so respectful of peace as the one in which Western democracies live and breathe, the truth is that outside these walls there are threats, and there are risks to the continuity of any particular nation.

The world remains a volatile, uncertain, and dangerous place with states, transnational organizations, and non-state actors all working in their own self-interests—which may or may not be aligned with the national interest of the Western world. There is little choice for Western countries but to maintain an active role to counter, impede, and dissuade hostile states, non-state actors, and transnational criminal organizations. Working in concert with other nations, international organizations, and Non-Governmental Organisations (NGOs), we must mitigate threats through the use of all elements of national power and focus on rule of law development as a means to provide stability.

If combat operations are required to wrest control of a nation or a large geographic area from a hostile force or if a nation becomes a failed state requiring international intervention, each participating nation must plan for and be prepared to implement programs to provide security and stability. Using as much of the indigenous criminal justice system as soon as possible should protect the people from harm and help them to begin developing a sense of "nation" to form the core of the nation that will rise from the ashes of the conflict.

Rule of law development requires a whole-of-government approach in which synchronization and coordination among the military, Embassy teams, international organizations, and NGOs are critical. In an operation with a kinetic component, or where the security situation may be unstable, the military must take the lead in developing the security umbrella using the criminal justice system for counterinsurgency and providing general security for the people. Other rules of law programs, focused on more generalized development efforts, have a longer time horizon and can more effectively flourish after the security situation is more stable.

For those who have deployed to or are familiar with war-torn areas or failed states, it is clear that military force alone will not be able to establish or implement rule of law. Legal systems and institutions take years to develop based on a variety of factors, including host nation culture, religion, and tolerated levels of corruption, and whether coalition members bring with them an ethnocentric bias that could complicate the establishment of rule of law. In at-risk, failed, emerging, and postconflict states, the military can set the conditions for rule of law development and stability by focusing on the state's criminal justice system.

The situation is not the same as it was fifty, thirty, not even ten years ago. That is why the current global security landscape is part of the major threats that can affect the West. Global security is affected by global threats like the ones posed by China, Russia, Iran, and North Korea but also and much closer to us, like the ones represented by Violent Extremist Organizations (VEO) and some rogue states or unstable ones like the countries we have in parts of our southern frontier.

We have seen this 2021 a clear and present danger to our sovereignty, and we have tested how difficult is the life on the vallado as Colonel Jessup once stated in the movie "A few good men". The threat analysis would be better and the results easier if the military were heard by our political masters, but unfortunately this is not the usual case. I guess we can blame some of the reasons behind this mistrust between civilian and military relations on history, tradition and on the wrong approach to discipline that affects some military leaders. Sometimes we see them or some of them trying to know what their bosses (no matter if military or civilian) want before submitting their proposals. If military leaders try to say what the politicians want to hear, being politically correct and not professionals, it will be difficult to defend a military point of view consistent with the mission given to the Armed Forces and the analysis, I was advocating to will be flaw. In military life, it is and will always be very difficult to concur in everything with our political masters, and this is why General Marshall trying to explain how he handled his somehow difficult relationship with the US political leadership stated: "I have always provided blunt and candid advice, I have always tried to keep my disagreements in a close hold environment, and I have always implemented faithfully decisions that went against mine". This, personally, is what I tried to follow in my own career, particularly in my tour of duty as Chief of Defense Spain.

But there is one more important thing, the portion referred to ethical behavior.

If there is something I am positive about is that in our quest for legitimacy, we need to have combatants with superior ethical behavior. We need soldiers able and willing to perform their duties respecting the law but warriors anyway. Warriors invested in a true fighting spirit, able to engage in combat when their own life is at risk. We need to have and to train soldiers inspired by a warrior ethos, people with a clear compromise and willing to risk their lives, to fight for a higher goal.

Because the second thing I am positive about is that the next generation or, at best, the following one will have to call its Armed Forces into action. At that very same moment, when the roll call starts, we cannot find our Armed Forces well prepared to fight fires or to stabilize more or less distant countries, and we need them prepared for their true mission, to guarantee our sovereignty and independence and so, able to fight for its defense and willing to keep intact its territorial integrity and constitutional order.

Please bear in mind that this lack of proper training and preparation has happened before. Twenty years ago, the Spanish Government decided to answer the call of one of its NATO allies shortly after the terrible events of 9/11. But the "war on terror" caught the Spanish Armed Forces ill-prepared, and we never got the opportunity to take part in the actual fighting that took place in Afghanistan or in Iraq afterward.

There were some reasons behind that lack of preparation that prevented real integration of our units in the allied formations. One of them was the lack of popular support in which the Spanish Government back then blamed almost everything. But the truth was that we were just in the middle of the process to professionalize our Armed Forces and we did not have units at a proper readiness status. The second one was related to the fall of the iron curtain at the end of the twentieth century. For some years, for too long, our units had been immersed back then in "peace operations" in the Balkans and elsewhere and, somehow, had forgotten to prepare for war. We had confirmation of this aspect when only a few years later we had to ask for close air support on route lithium in the Afghan province of Baghdis while our soldiers looked for cover from Taliban fire. The third and in my view more important constraint was that while we were planning for deployments of our troops, the Spanish Government was discussing if Armed Forces must fight a war abroad and really concerned about the so-called zero-losses policy and the impact of body bags in the media. "Zero-losses" policy permeated our units, and the commanders came to the conclusion that there could be no casualties at war. Of course, caveats and restrictions to their freedom of movement impacted the combat morale and affected heavily their warrior ethos.

The simple fact of knowing that they could not fire back insurgents if they were actually not firing over them made me, at that time Chief Ops at the Army HQ, to read some reports saying "after taliban attack, they concealed their weapons, and we were forced to let them flee" adding, "insurgents will harass us again tomorrow morning". There is no need to tell how difficult it was to take part in the Iraqi Task Force withdrawal in 2004, but it was even more difficult to assume, due to the unpopularity of the Afghan campaign and the losses over there, that our units were to play a second-line role in the counterinsurgency effort. No need to talk about the impact of such a policy on the morale of young officers and NCOs. Many of those who deployed there had the impression that their bosses intended to "go to Bagdhis, avoid trouble and return home as soon as possible" that it was "better not to exit the base in order to avoid risks".

And so we lost part of the ethics of the fighter, parts of the "warrior ethos" when we came to think that "the mission was not worth risking our lives". There is another input to this lack of ethos; it is related to a secondary task given to militaries across the world, the task of supporting civilian populations when the need arises. Supporting our societies when they are hit by a pandemic or a huge snowstorm is part of the Armed Forces nature but is only a secondary task.

If soldiers confirm that their own population does not support them when they are sent to fight a war, but at the same time this population shows a lot of support and caring for them when they are tasked to take care of elder people or to provide shelter from the storm, they will have an inclination to look for more rewarding jobs than patrolling in the desert.

And it will be almost impossible to maintain that we are warriors made for war. It will be difficult to explain our societies and political masters that we are not supposed to clean the streets, we are not trained to lift debris, and we do not plan to hand out colored pencils. It will be difficult to state that we are warriors, technically, physically, and psychologically trained for war that we can engage in streets, debris, nursing homes or colored pencils when the nation has no other instance to rely on but always keeping in mind this is not our primary task.

In conclusion, as long as our societies and our rulers find difficult to accept that the Armed Forces have to know how to make and win any possible war, little can be done. As long as it is not understood that a half-victory or a tie is not enough in matters of national security, we will not be able to have the Armed Forces that Western societies deserve and need. Until our societies are aware of the need to give legitimacy to its Armed Forces and to recover that warrior spirit in our military lives accepting combat as something substantial of our profession, our Armed Forces will continue to fight for an almost impossible goal and the risks that Western societies will have to accept will be, in my opinion, too high.

Four Stars General Fernando Alejandre Martínez
22nd Chief of the Defence Staff in the Kingdom of Spain (2017–2020)
Madrid, Spain

Introduction

We conceived the present book entitled *Ethical and Legal Challenges in the Face of Current Conflicts*, after the development of the LVII CICA International Conference and as a consequence of one of its conclusions: The conflicts of the XXI century must be approached from a perspective multidisciplinary and maintaining a solid ethical approach. Otherwise, the solutions to the challenges could be even more harmful than the problems to be solved.

For this reason, the CICA International Foundation, with the priceless sponsorship of the Antonio de Nebrija University and the always generous financing of Banco de Santander, launched the call to create a book that collects the opinion of experts from different scientific fields regarding what they will consider the main conflicts of the moment. The result of this call and the selection process of authors and themes is the book the reader has in his hands. As can be understood when approaching the table of contents, the topics addressed are very heterogeneous, as is the reality of postmodernity in which we so frequently find contradictory phenomena, of a complex nature, with many edges that can be approached from very different perspectives.

And that is why the selection of the authors is also heterogeneous, counting among them military personnel of the highest rank along with others of lesser rank; civilians from the university environment together with business leaders from the defense sector or academics from the field of philosophy; young authors along with others with a long career in the field of research. In the same way, the authors' area of specialization is heterogeneous, among which we find engineers, lawyers, philosophers, soldiers, doctors, internationalists, psychologists, or journalists. All this tries to balance the different perspectives and experiences that allow treating the chosen topics from points of view that are mutually enriching.

The themes have been grouped into two parts or sections. The first of them is dedicated to science and technology, with a particular emphasis on the cybernetic field, which is one of the fields in which current conflicts are unfolding powerfully; the second, with a greater depth and relationship with the ethical sphere, addresses problems and conflicts of our time in which the weight of technology is more petite.

In the first part, we begin with Dr. Federico Yaniz Velasco's contribution to the problems that are increasingly present concerning outer space, its use and use by powers, and nascent private companies. And on space as well, but regarding its increasing militarization, it is followed by the contribution of the prestigious Prof. Neuneck, President of the Pugwash movement in Germany, who addresses the problem that a military space field of operations poses for all humanity. It is followed in the expository order by the chapters dedicated to the cyber world, starting with the contribution of the members of the Colombian War School, Dr. Navas-Camargo and Colonel Ardila, who echo the proposals on the topic of Latin American Bank for Development; next, Dr. Bartolomé, an academic at the Inter-American Defense College in Washington, analyzes the prominent cases of security breaches that have occurred in this second decade of the twenty-first century, as well as Commissioner of the National Police Corps and today the business leader of the private sector of cybersecurity in Spain Bernardino Cortijo brings us closer to one of the possible solutions for corporations to assume the risks of the cyber environment and that is to follow cyber compliance policies. Three academic profiles culminate this first part, one consecrated as that of Prof. Pilar Otero González, who delves into the procedural problems of the prosecution of computer crimes; another in the full development of its capacity, that of Prof. Méndez Rocasolano, who analyzes the links and risks for human rights derived from the use of technologies such as artificial intelligence or big data; and finally one in training, that of Pablo Moral Martín, who enters fully into the problem of misinformation and fake news as a challenge for national security.

The second part presents a more extraordinary thematic richness. It encompasses very diverse conflictual problems and without a common link between them beyond that of posing risks to the security and stability of states. Dr. Farnicka focuses on modern psychology, her specialty, and the connection between this area of knowledge and conflict. It is followed by the contribution of the person who writes these introductory lines, seeking to delve into the philosophical reasons for the growing conflict in terms of territorial unity in some European states, with particular reference to the case of Spain. The member of the Navy, Mr. Juan del Pozo, contributes his ethical vision of the field of military operations and the necessary link of war actions to international humanitarian law and respect for human rights. And it is followed in the expository order chosen by Prof. Regí, who analyzes in detail the different solutions provided by three very different models of facing the fight against the pandemic that has severely impacted the world and that compromises the development of the decade for the economic derivatives that it has entailed. Another of the most pressing problems in Western European society is energy risk, which is discussed in our book by the Prof. Dr. Agnieszka and Prof. Piwowarsky. And our book ends with two chapters dedicated to another of the significant challenges of our society, which is to keep freedom of expression intact under challenging times. From a more theoretical framework, Dr. Boulos, and a more applied perspective, Professors Moreno Mercado, Calatrava, and Calvillo address this issue with the necessary academic depth.

The methodologies followed by the authors in the different chapters are those of their respective scientific fields. We think they enrich the non-specialized reader by offering a reasonably complete overview of areas of knowledge and experiences, which was the object of this book from the beginning.

Dr. Juan Cayón Peña

Contents

Ethics and Law Facing Other Current Conflicts

Sciences and Technologies; Cyberworld Conflicts

Outer Space's Legal Framework, Challenges, and Policies

Federico Yaniz

Abstract The peaceful use of space has opened new possibilities for human progress. The first artificial satellite on our planet, Sputnik 1, was put into orbit on October 14, 1957. The US National Aeronautics and Space Administration (NASA) opened for business on October 1, 1958, and since then has been one of the players in the conquest of space. For many years the Agency followed the same process for spacecraft development and operations. That process proved successful for more than 50 years. In 2005, mandated by President George W. Bush on January 14, 2004, NASA completed a successful year of milestones and discoveries starting the implementation of the Vision for Space Exploration, the USA long-term plan for returning astronauts to the Moon to prepare for voyages to other destinations in the solar system. The commercial space market leveled out also in 2005, and SpaceX announced plans to pursue a human-rated commercial space program through the end of the decade. That program would later become the Dragon spacecraft.[i] In 2006, NASA announced that the company was one of two selected to provide crew and cargo resupply demonstration contracts to the International Space Station (ISS) under the COTS program.[ii] Of 52 successful launches worldwide in 2005, about 23 (44%) were commercial launches compared to 19 (36%) in 2004. The situation had changed, and NASA had to adapt to a new environment and the privatization of the space industry. As a matter of fact, in the late 1960s, the US Government invested over 4.5% of the Federal Budget in NASA, and in 2005 the percentage was 0.63% as the US Government had to cut funding for many of NASA projects to reduce the deficit to boost the economy. The privatization of outer space exploration is a matter heavily debated in the international community. This paper will be studying the differences between the traditional NASA program and the more "open" model used today for the Commercial Crew and Cargo programs. Outer space is a neutral territory designated free from the appropriation under international law. Nevertheless, there is an approach towards using the space for defensive purposes that some could consider militarization.

Keywords Outer space · EU · NATO · Capsule · NASA · Operational domain

F. Yaniz (✉)
Eurodefense, Madrid, Spain

J. Cayón Peña (ed.), *Security and Defence: Ethical and Legal Challenges in the Face of Current Conflicts*, Advanced Sciences and Technologies for Security Applications,
https://doi.org/10.1007/978-3-030-95939-5_1

1 Foreword

Outer space has become an essential element for the development of Humanity. Indeed, the peaceful use of space for all kinds of activities provides many new possibilities for human progress. *Sputnik 1* was in orbit on October 14, 1957. Significant progress has been in the conquest of space since then.

There is not a generally agreed definition of the border between the Earth's atmosphere and outer space. However, the International Aeronautical Federation has established the Karman line at an altitude of 100 km as a practical working boundary between the atmosphere and outer space. In recent years, outer space has become an essential element for the further development of Humanity. The first steps of the exploration of space were slow: Yuri Gagarin was the first human sent into orbit on April 12, 1961, and only after eight years, on July 20, 1969, U.S. astronauts Neil Armstrong and Buzz Aldrin landed onboard of the Apollo Lunar Module Eagle on the Moon. Six hours later, Neil Armstrong was the first man walking on the Moon accomplishing one of the oldest dreams of human civilizations. (Michael Collins[iii] was flying the Apollo 11 command module Columbia around the Moon while his crewmates, Armstrong, and Aldrin, were landing the Apollo Lunar Module Eagle on the surface of the Moon.)

The number of artifacts that populate the near space environment has rapidly increased in the past years as space junk multiplied. The Orbital Debris Quarterly news,[iv] on Volume 25, Issue 2, June 2021, informs that as of April 4, 2021, there were 15.457 Spent Rocket Bodies and Other Cataloged Debris in space.

Also very relevant is the number of space stations that provide habitats for humans in space on longer-duration missions. These stations do not possess abilities to launch or land. Several space stations have orbited the Earth since 1971, with two currently active. This new stage in the conquest of space began when Salyut 1 was in orbit on April 19, 1971. According to Zarya.info,[v] a total of 9 Salyut space stations have orbited Earth: 5 DOS-type civilians and 4 Almaz-type military personnel. Skylab was the first United States space station. Launched by NASA and occupied for about 24 weeks between May 1973 and February 1974. It was operated by three separate three-astronaut crews: Skylab 2, Skylab 3, and Skylab 4. Relevant operations included an orbital workshop, a solar observatory, Earth observation, and hundreds of experiments.

Shenzhou V spacecraft launched on October 15, 2003, with the first Chinese man in space. Tiangong 1 (Celestial Palace 1) was China's prototype space station, and it was launched un-crewed aboard a Long March 2F/G rocket. Tiangong 1 was the first operational component of the Tiangong program. It orbited the Earth from September 29, 2011, till April 2, 2018, and received visits by a series of Shenzhou spacecraft during its two-year operational lifetime. The crewed Shenzhou 9 mission docked in June 2012. With the Tiangong program, China aims to place a modular station into orbit by 2023.

The International Space Station (ISS) was first launched in space by Russia on November 20, 1998, with the first humans successfully working in the station from the year 2000. Since then, there have been astronauts working continuously on the ISS. It is currently operating and permanently inhabited. The ISS is the product of the partnership Roscosmos, NASA, the Canadian Space Agency, the Japanese Aerospace Exploration Agency, and the European Space Agency. Two main mission control centers are being in Moscow and Houston.

As of July 20, 2021, there are ten people in space.[vi]

2 Legal Frameworks of Activities in Outer Space

Five core treaties regulate outer space activities: Treaty on Principles Governing the Activities of States in the Exploration and Use of Outer Space, including the Moon and Other Celestial Bodies or Outer Space Treaty (1967); The Rescue Agreement (1968); The Liability Convention (1972); The Registration Convention (1976); and the Moon Agreement (1984). Article IV of the Outer Space Treaty (1967) (UN, 1967) explains well the purpose of this group of treaties.

> Article IV. States Parties to the Treaty undertake not to place in orbit around the earth any objects carrying nuclear weapons or any other kinds of weapons of mass destruction, install such weapons on celestial bodies, or station such weapons in outer space in any other manner. The moon and other celestial bodies shall be used by all States Parties to the Treaty exclusively for peaceful purposes. The establishment of military bases, installations and fortifications, the testing of any type of weapons and the conduct of military maneuvers on celestial bodies shall be forbidden. The use of military personnel for scientific research or for any other peaceful purposes shall not be prohibited. The use of any equipment or facility necessary for peaceful exploration of the Moon and other celestial bodies shall also not be prohibited.
>
> Article IV The Outer Space Treaty. (UN, 1967)

There are other documents that complete the above-mentioned core treaties. The United Nations prepared the Outer Space Treaty recognizing the "legal problems which may arise in the carrying out of programs to explore outer space." On December 12, 1958, by Resolution 1348 (XIII), the United Nations General Assembly (d) (UNGA, 1958) designated outer space free from appropriation and limited for "peaceful uses." Additionally, the Outer Space Treaty bans nuclear weapons from outer space but does not explicitly regulate conventional weapons. Space Laws based on freedom cannot be shared for as many as possible without limitations. In any case, outer space is a very touchy issue when we consider the militarization of space. The Treaty Banning Nuclear Weapons Tests in the Atmosphere, in Outer Space and Under Water known as Partial Test Ban Treaty (PTBT, 1963), was signed by the governments of the Soviet Union, United Kingdom, and the United States of America in Moscow on August 5, 1963, before being opened for signature by other countries. The PTBT formally went into effect on October 10,

1963. Since then, one hundred and twenty other states have become parties and ten states have signed but not ratified the PTBT.

The UN General Assembly Resolution 1884 (XVIII) that was adopted unanimously on October 17, 1963, (UNGA, 1963), calls upon States to refrain from placing in orbit around the Earth any objects carrying nuclear weapons or any other weapons of mass destruction or from installing such weapons on celestial bodies. It is convenient to distinguish the legal framework for the whole outer space and the limits concerning the Moon and other spatial bodies. Article IV of the Outer Space Treaty provides that: State Parties to the Treaty undertake not to place in orbit around the Earth any object carrying nuclear weapons or any other kinds of weapons of mass destruction, install such weapons on celestial bodies, or station such weapons in outer space in any other manner. The Moon is the celestial body closest to Earth. In December 1979, the Moon agreement was signed (UN, 1979), following an initiative by the Soviet Union. On December 5, 1979, the UN General Assembly adopted the Agreement in resolution 34/68. The Moon Treaty supplements the Outer Space Treaty confirming the demilitarization of the Moon and other celestial bodies as provided for in that treaty but neither the US or Russia nor many European countries, Germany, and Spain among them, have signed the Moon Treaty.

> Article 3. States Parties shall not place in orbit around or other trajectory to or around the Moon objects carrying nuclear weapons or any other kinds of weapons of mass destruction or place or use such weapons on or in the Moon. The establishment of military bases, installations and fortifications, the testing of any type of weapons and the conduct of military maneuvers on the Moon shall be forbidden...
>
> Article 3 The Moon Agreement of 1979. (UN, 1979)

The word "peaceful" is used in treaties, resolutions, and conventions related to outer space. Some experts consider peaceful something that is non-aggressive and refers to the need to retain the right of self-defense, as expressed both in customary law and in Chapter VII Article 51 of the Charter of the United Nations (UN, 1945). Article 51 concerning "Action with respect to Threats to the Peace, Breaches of the Peace, and Acts of Aggression" states:

> 51. Nothing in the present Charter shall impair the inherent right of individual or collective self-defense if an armed attack occurs against a Member of the United Nations, until the Security Council has taken measures necessary to maintain international peace and security. Measures taken by Members in the exercise of this right of self-defense shall be immediately reported to the Security Council ...
>
> Article 51, Charter of the UN. (UN, 1945)

To preserve the sustainability of space activities is a challenging task. There is a need to revise and reform the treaties and reach a general agreement to avoid the existing status quo in space falls apart.

3 UN Space Agencies

From the earliest days of space exploration, the United Nations recognized the relevant role that space-related technologies can play in improving the human condition throughout the globe. To this end, the United Nations and its specialized agencies conduct activities utilizing these technologies. Five treaties and several UN General Assembly resolutions regulate that all areas of outer space are devoid of weapons of mass destruction, whether for storage, experimentation, or use. On the other hand, the Moon, and other celestial bodies, generally exclude military activity. This conclusion makes it possible to develop or envisage certain military activities, in particular defensive actions, in outer space having in consideration the right of self-defense as in Article VII, Chapter VII Article 51 of the Charter of the United Nations (UN, 1945).

The UN has two relevant bodies related to the activities in outer space: the Committee on the Peaceful Uses of Outer Space (COPUOS) and the Office for Outer Space Affairs (OOSA), known as UNOOSA. COPUOS, set up by the General Assembly in 1959, governs the exploration and use of space for the benefit of all humanity: for peace, security, and development. COPUOS oversees reviewing international cooperation in peaceful uses of outer space, studying space-related activities that could be undertaken by the United Nations, encouraging space research programs, and studying legal problems arising from the exploration of outer space. The Committee was instrumental in the creation of the five treaties and five principles of outer space. International cooperation in space exploration and space technology applications to meet global development goals are discussed in the Committee every year. Space activities are constantly evolving, owing to rapid advances in space technology. The Committee, therefore, provides a unique platform at the global level to monitor and discuss these developments.

UNOOSA is an office of the General Assembly of the United Nations responsible for promoting international cooperation in the peaceful uses of outer space. The Office forms part of the United Nations Office at Vienna and is the Secretariat for the General Assembly's Committee on the Peaceful Uses of Outer Space. The Committee has two subcommittees: the Scientific and Technical Subcommittee and the Legal Subcommittee. The Office for Outer Space Affairs implements the United Nations Program on Space Applications (PSA), improving the space science and technology for the economic and social development of all countries, particularly developing countries. Under the Program, the Office conducts training courses, workshops, seminars and, other activities in subject areas such as basic space science, satellite navigation, etc. On behalf of the UN Secretary-General, the UNOOSA maintains the Register of Objects Launched into Outer Space and disseminates via its website the information recorded. Since 1962, the United Nations maintains a Register of Objects Launched into Outer Space. Initially established as a mechanism to aid COPUOS, the evolution of international space law resulted in space object registration becoming a means of identifying which states bear responsibility and liability for space objects.

Following multi-year discussions among States, the Convention on Registration of Objects Launched into Outer Space entered into force in 1976. States and international intergovernmental organizations that agree to abide by the Convention are required to establish their national registries and provide information on their space objects to the Secretary-General for inclusion in the United Nations Register.[vii] Delegated by the Secretary-General, UNOOSA has the responsibility for maintaining the Register and disseminates the information provided as United Nations documents. That information is available through its website and the United Nations Official Document System. Over 88% of all satellites, landers, crewed spacecraft, and space station flight elements launched into Earth orbit or beyond are in the Register.

4 Space Exploration and Private Industry

In the United States, there is renewed interest in outer space. In December of 2017, President Trump signed Space Policy Directive-1, in which he directed NASA: "to lead an innovative and sustainable program of exploration with commercial and international partners to enable human expansion across the solar system and to bring back to Earth knowledge and opportunities."

Following that mandate, on April 16, 2018, President Donald Trump signed the document NASA's Exploration Campaign: Back to the Moon and on to Mars (NASA, 2018). He said:

> The directive I am signing today will refocus America's space program on human exploration and discovery. It marks a first step in returning American astronauts to the Moon for the first time since 1972, for long-term exploration and use. This time, we will not only plant our flag and leave our footprints, but we will also establish a foundation for an eventual mission to Mars, and perhaps someday, worlds beyond.

Apollo 17 was the last mission to the Moon in 1972. This NASA mission took Eugene A. Cernan and Harrison H. Schmitt[viii] to the lunar surface and marked the end of the Apollo program. By the time of the launch of Apollo 17, on December 7, 1972, public excitement in space exploration had weakened earlier, and the US Government had turned the attention to Vietnam. Those and numerous other factors led the Apollo program to a close.

The new National Space Exploration Campaign (NSEC) calls for human and robotic exploration missions to expand the frontiers of human experience and to discover natural phenomena of Earth, other worlds, and the cosmos. The construction of NASA's Space Launch System (SLS) and Starliner reusable crew capsule will provide the foundation for exploration beyond Earth's orbit. The Boeing CST-100 (Crew Space Transportation-100) Starliner is manufactured by Boeing as its participation in NASA's Commercial Crew Development (CCDev) program. Its primary purpose is to transport crew to the International Space Station (ISS) and private space stations such as the proposed Bigelow Aerospace Commercial Space Station. The NSEC does not assume or require significant funding increases and builds on 18 years of astronauts and cosmonauts living and working together on the ISS. Another sign of the USA's renewed interest in space was the already mentioned restoration of the Space Command as a unified combatant command, announced at an event presided over by President Trump on August 29, 2019. Charles Beames, executive chair of York Space Systems and chair of the SmallSat Alliance said about the position of Biden Administration on space:

> Looking ahead, the decisions the current administration must make regarding the Council, the Space Force, NASA and commercial space policies will determine whether space can remain a safe, nonpartisan domain for an economy to flourish or become an inhospitable orbital minefield where only military hegemons joust for supremacy.

4.1 NASA Space Exploration

The refurbished NASA has a plan for yearly Moon flights through 2030. The first flight of NASA's SLS will carry astronauts into deep space and return them to Earth. NASA's Orion spacecraft is prepared to take humans farther than they have ever gone before. Orion will serve as the exploration vehicle that will carry the crew to space, provide emergency abort capability, sustains the astronauts during the space travel, and provide safe re-entry from deep space return velocities. This spacecraft will take the crew to the "lunar Gateway" in a lunar orbit crew capsule. The lunar Gateway, or simply Gateway, is a small space station in lunar orbit intended to serve as a solar-powered communication hub, science laboratory, short-term habitation module, and holding area for rovers and other robots. Formerly known as the Deep Space Gateway (DSG) and renamed Lunar Orbital Platform-Gateway (LOP-G) will fly for the first time in 2024. Together, Orion, SLS, and the LOP-G represent the basis of NASA's sustainable infrastructure for the next step in human exploration. The LOP-G will enable crew expeditions down to the Moon surface, facilitating the exploration of new locations across the Moon. The Gateway can operate autonomously and automatically as a deep space science outpost even without crew. Over time, the LOP-G will become a way station for refueling depots, servicing platforms, and a facility for processing samples from the Moon and other celestial bodies.

The SLS and Orion are critical to NASA exploration plans at the Moon and beyond. The SLS is the most powerful rocket for sending humans on missions to deep space. Orion spacecraft can maintain humans alive hundreds of thousands of miles from home. Exploration Flight Test-1 (EFT-1) was the first test of the Orion Multi-Purpose Crew Vehicle. Without a crew, it was launched at Cape Cañaveral Air Force Station on December 5, 2014, at 12:05 UTC (7:05 am EST), by a Delta IV Heavy rocket from Space Launch Complex 37B. The Delta IV Heavy (Delta 9250H) is an expendable heavy-lift launch vehicle, the second highest-capacity rocket in operation, and the largest of the Delta IV family. Manufactured by United Launch Alliance and was first launched on December 21, 2004. Designed originally by Boeing's Defense, Space, and Security division for the Evolved Expendable Launch Vehicle (EELV) program, the Delta IV became a United Launch Alliance (ULA) product in 2006.

4.2 Private Industry in Space

The participation of Starliner and private companies such as SpaceX and Blue Origin are changing the future of space exploration and travel. Space Exploration Technologies Corporation, known as SpaceX, designs, manufactures, and launches advanced rockets and spacecraft. The company, founded by Elon Musk in 2002, aims to revolutionize space technology, reduce space transportation costs to Mars, and enable people to live on that and other planets. SpaceX manufactures the Falcon 9 and Falcon Heavy launch vehicles family, several rocket engines, Dragon cargo/crew spacecraft, and the Starlink satellites. In 2010, SpaceX changed the standards in space exploration when it became the first private industry to launch a spacecraft into orbit. It was also the first private company to send a spacecraft to the ISS and the first to send astronauts into space. The Chief Executive Officer (CEO) Elon Musk and the Chief Operating Officer (COO) Gwynne Shotwell have declared that SpaceX plans to put astronauts on Mars.

Blue Origin is also a privately-owned aerospace company founded by Jeff Bezos in 2000 to enable millions of people to work in space. Blue Origin is developing a variety of technologies, with a focus on rocket-powered vertical takeoff and vertical landing (VTVL) vehicles for access to suborbital and orbital space. Initially focused on suborbital spaceflight, the company has designed, built, and flown multiple testbeds of its New Shepard vehicle at its facilities in Culberson County, Texas. Test flights of the New Shepard, named after the first American in space Alan Shepard, began in April 2015 and eventually successfully flew its first crewed mission on July 20, 2021.

SpaceX's achievements include being the first privately funded liquid-propellant rocket to reach orbit (Falcon 1 in 2008) and the first private company to launch, put in orbit, and recover a spacecraft (Dragon in 2010). Furthermore, SpaceX was the first private company to send a spacecraft to the ISS (Dragon in 2012) and the first vertical take-off and vertical propulsive landing for an orbital rocket (Falcon 9 in 2015). Other achievements have been the first reuse of an orbital rocket (Falcon 9 in 2017) and the first private company to send astronauts to orbit and to the ISS: SpaceX Crew Dragon Demo-2 and SpaceX Crew-1 missions in 2020. SpaceX has flown the Falcon 9 series of rockets over one hundred times. Furthermore, SpaceX is developing a large Internet satellite constellation named Starlink. Reusability is an integral part of the Falcon program. SpaceX pioneered reusability with the first re-flight of an orbital class rocket in 2017. As already mentioned, the company sent its first two astronauts to the ISS on May 30, 2020, aboard the SpaceX Crew Dragon. A few months later, on November 15, 2020, the successful launch of SpaceX Crew-1 with four astronauts took place. As of early 2021, it is the only commercial spaceflight company capable of sending astronauts to space, although it may soon face competition from Boeing's CST-100 Starliner.

On April 28, 2021, a SpaceX Falcon 9 rocket launched a new batch of 60 Starlink Internet satellites into orbit and nailed a landing at sea to top off a successful mission.

The veteran Falcon 9 rocket blasted off from Space Launch Complex 40 at Cape Cañaveral Space Force Station in Florida on the company's 10th launch of the year.

SpaceX Crew Dragon capsule carrying four astronauts to the ISS, as part of the company's Crew-2 mission, docked successfully at the orbiting ISS early Saturday, April 24, 2021. SpaceX's Crew Dragon capsule and its crew of four undock from the ISS on April 30, 2021, and splashdown just before midday on Saturday, May 1, 2021. Strapped inside the Dragon capsule, called Resilience, four astronauts[ix] made the first US night water landing in more than 50 years.

SpaceX is one private company that demonstrates the benefits of competition, efficiency, innovation, and collaboration between the public and private sectors. But in any case, the privatization of outer space exploration is heavily debated in the international community. Many developed countries support privatization, as they are the base for companies such as SpaceX, Magellan Aerospace, Blue Origin, Spaceship One, Virgin Galactic, and more. One advantage in utilizing private companies is that they take more risks which usually leads to innovation. A report by Business Insider[x] found that the crew risk number is 1 in 276, and Musk reported that Falcon Heavy had a 50% chance of blowing up, but he took that risk anyway. These developments will continue, but failures are possible.

5 Challenges of Privatization

The significant growth of the private space industry makes it necessary to answer the question of whether the privatization of the exploration of outer space is sound and convenient for the future of Humanity. As have been presented in Chapter 2 there are laws and regulations that are important to recognize in terms of territory and exploration. COPUOS has subcommittees that focus on this fair treatment. There is always a need for more indemnification policies and clauses to be added to these treaties, but they are a key tool in space exploration.

Private space industry companies help humanity to explore living conditions on other planets, develop new understandings of outer space, and increase our standard of living. The success of SpaceX's most recent launch and its accessibility to the ISS will begin to change the relations when it comes to sending astronauts to space. Since 2011, the USA and Russia have formed a mutually beneficial relationship between NASA needing to send astronauts to the ISS and Russia benefiting from NASA's money. Now SpaceX and NASA are working together, and the status quo may begin to change. Dmitry Rogozin, director of Russia's Roscosmos, has publicly congratulated NASA and SpaceX. However, the statistics show that Russia seems to have relied on NASA's money for its production and that 17% of their Roscosmos budget came from NASA as can be seen in the Space Threat Assessment Report 2020 (CIS, 2020). The shift towards even more significant privatization of space exploration in the US will have some impact on Roscosmos as a longtime partner of NASA.

We have seen some of the indisputable advantages of working with private companies. Industry privatization is commonly considered essential to capitalism and very useful in some industries. Nevertheless, there are also some critics of the privatization of space exploration.[xi] Private companies are supposed to obtain profit and must satisfy the interest of the shareholders. Meanwhile, public organizations such as NASA and the European Space Agency (ESA) must address the needs of the government and the taxpayers. The profit motive is always behind the goals of private enterprises; for that reason, they are to work in programs with rapid implementation of commercial applications. In a rapidly expanding sector as the aerospace industry, progressive nations should be encouraged to work with private companies to obtain maximum productivity. However, the public sector is paramount in the development and maintenance of some no commercial space programs. For that reason, public companies assure a balanced development of space exploration. In any case, it is a must to be careful to avoid uncontrolled competition that could potentially ignite a dangerous situation. The benefit of a regulated collaboration between public and private sectors would avoid possible unhealthy competition and make possible that exploration of space is a success for the future of Humanity.

Private industries are not only interested in the exploration of outer space but also other activities such as spatial tourism. We have already seen in 2021 some initiatives and very relevant new activities with the participation of private industries. For example, normalize tourism in space is a challenge for the private space industry and the regulatory authorities.

The participation of private industries in activities in space will continue in the foreseeable future. One example is the agreement between NASA and Texas-based company Axiom Space (NASA-Axiom Space Agreement, 2021) on terms for the first private astronaut mission to the ISS, which will launch as soon as January 2022. The agreement, announced on May 10, 2020, includes only a portion of the exchanges required to make a flight like this a reality, but it will result in a net payment from NASA to Axiom of $1.69 million. The agreement will allow Axiom to send a retired NASA astronaut and three passengers to the orbiting laboratory aboard a SpaceX Crew Dragon capsule for a journey of about a week in the first crewed space station mission for exclusively private interests. Phil McAlister, director of commercial spaceflight development at NASA, said in a news conference: "This is a real inflection point, I think, with human spaceflight."

On July 11, 2021, millionaire Richard Branson has successfully reached the edge of space onboard his Virgin Galactic VSS Unity spacecraft. Unity launched from the Eve mothership at an altitude of 50,000 ft. Branson flew high above New Mexico in the vehicle that his company had been developing for 17 years. He said the trip was the experience of a lifetime. They returned safely to Earth just over an hour after leaving the ground. The trip also makes them the first of the new space tourism pioneers to try out their vehicles, beating Blue Origin's Jeff Bezos. Richard Branson "beat" Bezos to space by nine days.[xii] Virgin Galactic already has more than 600 reservations at $250,000 apiece. Founded by Branson in 2004, the company has sent the crew into space four times and plans two more test flights from New Mexico before launching customers next year.

On July 20, 2021, millionaire Jeff Bezos made a short spatial trip in the first crewed flight of his rocket ship, New Shepard. Mark Bezos, his brother; Wally Funk, pilot and 82-year-old pioneer of the space race; and 18 years old student Oliver Daemen accompanied him. When the capsule touched back down after the 10 min and 10 s flight, Jeff Bezos exclaimed: "Best day ever!". The capsule touched down on the desert floor near Van Horn in remote West Texas after an automated 10-min flight. Blue Origin's New Shepard ship made its trip on the 52nd anniversary of the Apollo 11 moon landing, a date chosen by its historical significance. Bezos and his crew successfully crossed the Karman Line, an imaginary boundary considered by many the beginning of space. Unlike Branson's piloted rocket plane, Bezos' capsule was automated and required no official staff on board for the up-and-down flight.

6 Defense, Militarization and Space

Sputnik 1, the first artificial satellite of our planet, was put in orbit on 14 October 1957. That launch from the Baikonur Cosmodrome (now in Kazakhstan) ushered in new political, military, technological, and scientific developments. It also marked the start of the space age and triggered the U.S.-U.S.S.R. space race. Less than four years later, Yuri A. Gagarin became the first man to enter space on 12 April 1961, when he made a flight that orbited Earth lasting one hour and 48 min in his Vostok 1 spacecraft. Yuri Gagarin had become a lieutenant of the Soviet Air Force on 5 November 1957, and, on 6 November 1959, he had got the rank of senior lieutenant of the Air Force of the USSR. Gagarin's flight was a triumph for the Soviet space program, and he became a national hero of the Soviet Union. Newspapers around the globe published his biography and details of his flight. Gagarin was in a long motorcade through the streets of Moscow and, Nikita Khrushchev awarded him the title Hero of the Soviet Union. Other cities in the Soviet Union also held mass demonstrations. Gagarin became lieutenant colonel of the Soviet Air Force on 12 June 1962. Then, he got the rank of Colonel of the Soviet Air Force on 6 November 1963. Soviet authorities tried to keep him away from any flights, worried about losing their hero in an accident. On 20 December 1963, Gagarin became Deputy Training Director of the Star City cosmonaut training facility. At the same time, he began to re-join his flights as a fighter pilot. On 27 March 1968, while on a routine training flight from Chkalovsky Air Base, he and flight instructor Vladimir Seryogin died in a MiG-15UTI crash near Kirzhach. In the Soviet Union it was considered from the beginning that activities related to outer space had a close relationship with security and defense.

More recently, some other countries are pursuing space programs not only for civilians but also for military purposes, following the example of the USA and the Soviet Union. After many years of discreet activities, in 2015, the Chinese People Liberation Army (PLA) established the Strategic Support Force, which handles space, cyber, and the electromagnetic spectrum. Russia also set up an independent Space Force in the same year. India conducted an anti-satellite weapon test in March 2019. In response to these developments, France also established the Space Command in September 2019. The United States Space Force (USSF), created on December 20, 2019, is the newest branch of the US Armed Forces; the United States Space Command (USSPACECOM) is the latest of the eleven unified commands in the US Department of Defense. Some other countries also invest in military satellites for reconnaissance. For example, Iran launched on April 22, 2020, the Noor (light), Iran's first military satellite. Simultaneously announced its parallel space program run by the Islamic Revolutionary Guard's Corps (IRGC).

For many years, the US military strategy in outer space has called for superiority which means the ability to exploit a territory (here space) while selectively disallowing it to adversaries. The United Nations has repeatedly declared that outer space is freely available for exploration and use by all. In 1967, the Outer Space Treaty designated outer space free from national appropriation and military operations.

7 The Atlantic Alliance's Approach to Space. A New NATO's Space Doctrine?

NATO is the most successful defensive alliance in history. Its main achievements are that after 72 years peace has been maintained in Europe and that the Cold War ended without all-out armed confrontation. Furthermore, many of the countries that were members of the Warsaw Pact are now active members of the Atlantic Alliance. Given the current strategic situation, NATO must be alert to the multi-directional risks that characterize today's world and to which they may appear in the future. For NATO traditional risks have been coupled with those involving terrorism, the proliferation of weapons of mass destruction, disruptive new technologies, failed states, and religious and political extremism. To the mentioned risks, we must add the others arising from possible attacks on the storage, processing, and transmission systems of data that work almost always in cyberspace. The proliferation of new threats and risk makes space increasingly important to NATO and to the security and prosperity of allies and partners.

Space capabilities by NATO members bring benefits to NATO in multiple areas from weather monitoring, nature, and agriculture, to transport, science, communications, and economy. Furthermore, information gathered and delivered through artificial satellites is critical for NATO activities: operations, and missions, including collective defense, crisis response, and counterterrorism. At the same time, outer space is becoming more crowded and competitive, and satellites are now more vulnerable to interference than ever before. Russia, China, and other countries have developed a range of counter-space technologies.

NATO is a forum for Allies to share information and coordinate activities on various space-related issues, and its approach to space will remain in line with international law. But it is worthwhile to remember that outer space is also essential to deterrence and defense. Outer space underpins the military ability of the Alliance: to navigate and track forces, have robust communications, detect missile launches, and ensure effective command and control. Therese Wood states in—Visualizing All of Earth's Satellites: Who Owns Our Orbit? (Woods, 2020) that there were in October 2020 more than 2.660 operational satellites plus 3.200 that could be considered junk. More than half of the satellites belong to NATO members or companies based on their territory. Space data, products, and services are critical enablers and directly support other operational domains. The evolution in the use of space and new advances in space technology have raised new opportunities, new potential threats, and vulnerabilities. Space is used for many peaceful activities but also for aggression by potential adversaries. Satellites can be hacked, jammed, or weaponized. Anti-satellite weapons (ASAT) could cripple communications and the capacity to operate. Anti-satellite weapons (ASAT) are space weapons designed to destroy satellites for strategic or tactical purposes. Some nations possess operational ASAT systems. Although no ASAT system has been present in a concrete warfare situation, few countries (United States, Russia, China, and India) have successfully shot down their own satellites to demonstrate their ASAT capabilities.

Russia, China, and other countries have developed a counter-space technologies that could restrict NATO access to and freedom to operate in space. Various risks to space systems are increasing and can harm the security of the Allies. The Alliance is not aiming to develop its space capabilities and will rely on national space assets. NATO's approach to space will remain fully in line with international law and has no intention to put weapons in space.

From a security and defense point of view, space is critical for NATO in positioning, navigation, and timing. To implement space as an operational domain, the Alliance is increasing its space domain awareness and the common understanding of the space environment, including threats and risks. Maintaining situational awareness and reliable access to space services are critical to ensure the success of operations, missions, and activities. At the 2018 Brussels Summit, NATO leaders recognized that space is a highly dynamic and rapidly evolving area, essential for the Alliance's security, and agreed to develop an overarching NATO Space Policy.

Space has facilitated NATO's performance of numerous tasks, functions, missions, and operations in peace and, if needed, at war. Some of these functions are essential and need support from space. Among them are the knowledge of the general situation and early warning, where the rapid identification of forces that may threaten allied borders is essential to deter or counter aggression. On the other hand, artificial satellites can detect missile launches and provide information on what other actors are doing on land, at sea, in the air, and space. The use of space has a significant impact on everyday life: communications, agriculture, weather forecasts, and broadcasting radio and television. Since the Brussels Summit of 2018, NATO's attention to space has been increasing dramatically. Point 19 of the Declaration issued on 11 July 2018 after that meeting of Heads of State and Government states:

> In the air domain, we have agreed a Joint Air Power Strategy, which is a key enabler for NATO's peacetime Air Policing and Ballistic Missile Defense missions. It will strengthen our Integrated Air and Missile Defense, and guide our aerospace capabilities to operate together jointly, more swiftly, and effectively in peacetime, crisis, and conflict… we have

> agreed to develop an overarching NATO Space Policy. NATO Brussels Summit Declaration, 11 July 2018. (NATO, 2018)

Space support to NATO is provided by national resources of allies or by commercial services distributed by NATO agencies. Therefore, a coordination among them is required to ensure that spatial support is continuous for the Alliance. The Space Support Coordination feature, located on some allied commands, channels that support. On June 27, 2019, Allied Defense Ministers approved the new allied space policy "(PO (2019)0279 (INV)), NATO Overarching Space Policy" (NATO, 2019a), that is the guide to the Alliance's activity in space. Five months later and after the meeting of the NAC in Foreign Ministers session on November 29, 2019, NATO Secretary-General J. Stoltenberg said:

> …We have agreed that space should be a new operational domain for NATO – alongside air, land, sea and cyber. Space is part of our daily life here on Earth. It can be used for peaceful purposes. But it can also be used aggressively…Space is also essential to the Alliance's deterrence and defense... And our approach will remain fully in line with international law…
>
> Stoltenberg, J. 20 November 2019. (Stoltenberg, 2019)

The London Declaration issued in December 2019 by NATO leaders states:

> To stay secure, we must look to the future together. We are addressing the breadth and scale of new technologies to maintain our technological edge, while preserving our values and norms… We have declared space an operational domain for NATO, recognizing its importance in keeping us safe and tackling security challenges, while upholding international law…
>
> NATO London Declaration, 4 December 2019. (NATO, 2019b)

The Alliance is advancing the integration of space in training and exercises, operational planning, capability development, and innovation efforts. Emerging technologies are transforming the space domain and, NATO is taking advantage of these developments to maintain its technological edge. In this regard, NATO's Science and Technology Organization network helps leverage the scientific capacity among Allies and partners.

To allow NATO forces to communicate more securely and quickly, NATO is investing over EUR 1 billion in procuring satellite communications services from 2020 to 2034. That is the biggest ever investment in satellite communications, provided by Allies and enabling more resilient and flexible communications with ships at sea, air assets, and troops across the globe.

On 22 October 2020, NATO Defense Ministers decided to establish a NATO Space Centre. Based on the Defense Ministers' decision, the NATO Space Centre will serve as a focal point to support NATO operations and missions, share information, and coordinate Allies' efforts. The Centre is building upon capabilities and personnel already at Allied Air Command in Ramstein, Germany. Over the next few years, the Centre will continue to grow, ensuring the required capacity across NATO and its military command structure.

In the Press Conference after the Ministers meeting on 22 October 2020 (Stoltenberg, 2020), the Secretary-General said:

> … NATO Defense Ministers have just met to address our deterrence and defense. NATO is determined to keep our cutting edge in all domains. Land, sea, air, cyber, and space... Ministers agreed to establish a new NATO Space Centre at Allied Air Command in Ramstein, Germany…
>
> Stoltenberg, 22 October 2020. (Stoltenberg, 2020)

On point 33 of the Brussels Summit Communiqué issued on 14 June 2021 it is stated:

> 33. We recognize the growing importance of space for the security and prosperity of our nations and for NATO's deterrence and defense. Secure access to space services, products, and capabilities is essential for the conduct of the Alliance's operations, missions, and activities. We will accelerate our work to deepen and expand our use of space as an operational domain, including through the NATO Space Centre in Germany and the upcoming establishment of the Space Centre of Excellence in France, which we welcome… Consistent with the Overarching Space Policy, NATO's approach to space will remain fully in line with international law…
>
> Brussels Summit Communiqué issued on 14 June 2021. (NATO, 2021)

When considering the possible militarization of space, NATO has made clear in many declarations and publications that its space approach will remain fully in line with international law. NATO has no intention to put weapons in space.

8 The European Union and Outer Space; The European Union Space Program

Outer space is used today for many civil and military applications, and it is an increasingly congested area. The European Union (EU) possesses world-class capabilities in space, such as Copernicus and Galileo systems, and highly competitive space industry. Furthermore, Europe's skills base in the space sector could be considered a capability. Moreover, work is ongoing on various projects, such as Governmental Satellite Communications and two projects under the Permanent Structured Cooperation (PESCO) framework: member states are developing an EU Radio Navigation Solution (EURAS) and a European Military Space Surveillance Awareness Network (EU-SSA-N). The services provided by EU space programs benefit millions of people. The EU has its space policy and programs, and they help with the implementation of EU policies. The most important are:

- Copernicus is a provider of Earth observation data.
- Galileo is Europe's global satellite navigation system.
- The European Geostationary Navigation Overlay Service (EGNOS) provides "safety of life" navigation services to aviation, maritime and land-based users over most of Europe.

EU space programs provide public services to EU authorities, companies, and citizens. Space data is essential to answering challenges such as the sustainable consumption of natural resources, safety and security, and climate change. Ensuring access to space assures the implementation of EU policies and the competitiveness of European industries and businesses, as its security, defense, strategy, and autonomy.

The Joint statement on a shared vision and goals for future Europe in space by the European Union and the European Space Agency Space Strategy (EU, 2016) signed on 26 October 2016, marks the objectives of the European Commission's Space Strategy for Europe:

- Maximize the benefits of space for society and the EU economy…
- Ensure a globally competitive and innovative European space sector…
- Reinforce Europe's autonomy in accessing space in a safe and secure environment…
- Strengthen Europe's role as a global actor and promoting international cooperation.

The emphasis on the economic dimension of space on the 2016 Space Strategy is easy to understand: the EU's space economy, including manufacturing and services, employs over 230.000 professionals, according to research carried out by the European Commission. Its value was estimated at around €50 billion in 2014, representing one-fifth of the value of the global space sector.

On 16 December 2020, the Council and the European Parliament reached a provisional political agreement on the Regulation of the European Parliament and the Council establishing the EU Space Program (EU, 2021). The financial envelope of €14.8 billion in current prices (€13.2 billion in 2018 prices) covers Galileo and EGNOS €9 billion and Copernicus €5.4 billion.

The agreement will ensure high-quality, up-to-date, and secure space-related data and services; socio-economic benefits from the use of such data and services, such as increased growth and job creation in the EU; enhanced security and autonomy of the EU; a more relevant role for the EU as a leading actor in the space sector.

The provisionally agreed text on 16 December 2020 was submitted to the Council's permanent representatives committee for analysis/political endorsement on Friday 18 December. However, the Regulation could not be adopted before the adoption of the EU's Multiannual Financial Framework for 2021–2027. In April 2021, the Council and European Parliament approved the Regulation establishing the new EU space program for 2021 to 2027. The program entered into force retroactively on 1 January 2021 (EU, 2021). Point 2 of the document explains why it is essential that the EU benefits from autonomous access to space:

> (2). The possibilities that space offers for the security of the Union and its Member States should be exploited as referred to in the EU Global Strategy for the Foreign and Security Policy (CFSP) of June 2016, while retaining the civil nature of the Union Space Program… Historically, the space sector's development has been linked to security… However, the Union's security and defense policy is determined within the framework of the CFSP, in accordance with Title V of the TUE.
>
> Regulation of the European Parliament and the Council establishing the (EU, 2021)

EU's economic power will suffer without access to space, a risk that emerges from the weaponization presented by Daniel Fiott in the report Securing the heavens, 15 April 2021 (Fiott, 2021). These risks are possible by the development of technologies such as Anti-Satellites weapons (ASAT). The Directorate-General for Defense Industry and Space (DEFIS), within the European Union Commission Internal Market Division, addresses four of the political priorities: a European Green Deal, a Europe fit for the digital age Supporting our European way of life, a Stronger Europe in the World, and the implementation of the Space Policy. The Space Policy includes the EU Space Program, ensuring access to space, EU Space Research, and Innovation initiatives, and investing in quantum technologies. The EU Space Program will promote the emergence of a European New Space eco-system to foster entrepreneurship and the European space industry.

The EU space components are in close cooperation with EU countries, the ESA, the European Organization for the Exploitation of Meteorological Satellites (EUMETSAT), the European Agency Global Navigation Satellite Systems (GNSS), and many other stakeholders. Ensuring effective and efficient cooperation and coordination between these actors is essential to optimize European policies and investment in space. Decision-makers, public authorities, EU citizens, EU commercial and private users, and NGOs are the main targets of the Program.

Many experts agree that the EU requires more investments in space capabilities and critical infrastructure protection. The Conclusions adopted by the Council of the EU at its meeting on 28 May 2019 on space as an enabler (EU, 2019), reflect the importance that the EU attaches to everything related to outer space. Other key documents of the EU related to space are the Convention establishing a European Space Agency of 30 May 1975 (EU, 1975) as well as the Framework Agreement between the European Community and the European Space Agency which entered into force in May 2004. This framework agreement calls for regular joint meetings of the Council of the European Union and the ESA Council at the ministerial level, known as the Space Council.

The European External Action Service (EEAS) is also involved in space policies promoting sustainable space operations. Mrs. Carine Claeys, Special Envoy for Space and Head of the EEAS Space Task Force, (Claeys, 2019), in a panel discussion at Euroconsult's World Satellite Business Week in Paris in September 2019, said:

> The Safety, Security and Sustainability of Outer Space (3SOS) public diplomacy (of the EEAS) initiative will promote "ethical conduct" in space amid concerns about orbital debris. (Claeys, 2019)

The belief that lows Earth orbits are becoming crowded with satellites and debris triggered the initiative 3SOS. The deployment of mega-constellations of satellites will aggravate the situation.

9 Final Remarks

Space exploration has advanced very much since the Nuclear Arms Race era. The exploration began after the Soviet Union's launch of *Sputnik 1* and was a game played only by the USSR and the USA for many years. Later, Europeans, Japanese, and others were able to join the race although with less ambition, more limited capabilities and at a much slower pace.

The participation of private companies is changing the future of space exploration and that participation will continue in the foreseeable future. One example is the agreement between NASA and Texas-based company Axiom Space on terms for the first private astronaut mission to the International Space Station (ISS), which will launch as soon as January 2022.

To advance in space activities, the EU is in favor of cooperation between private space industries and the European Space Agency, and other European public agencies.

Space debris is one of the principal threats to EU satellites and the main concern of the Union on space defense is to get a secured satellite-based infrastructure. The first challenge is to get an efficient space surveillance and tracking (SST) system that detects space debris, catalogs debris objects, and determines and predicts their orbits. Without an efficient SST system, the EU will not be able to prevent accidents or to establish early warning against intentional attacks.

On 11 November 2020, the EU Council adopted some important conclusions on "Orientations on the European contribution in establishing key principles for the global space economy" in preparation for the tenth Space Council[xiii] meeting, which was held by video conference on 20 November 2020. The text provides important orientations for future European space policy. In fact, the Council notes with satisfaction the emergence of a highly competitive European space industry and supply chains, which enables Europe to participate in the global growth of the space economy. With the purpose of fostering European space autonomy, security, and resilience, the mentioned Council on 11 November 2020 stressed the need for European technological non-dependence and recalled the importance of maintaining a secure, autonomous, reliable, cost-effective, and affordable access to space.

There are growing signs that space is going to be utilized for security as the concept of the battlefield has changed with advances in technology. New technologies to address air and missile threats are a clear example of this. The accuracy and reliability of ballistic missile defense (BMD) systems have improved in recent years, but developments in offensive technologies have outpaced this progress. China and Russia are developing flying object threats that overwhelm defensive reaction capability in Western countries. Hypersonic glide vehicles (HGV), for example, travel at Mach 5 or higher, and missile threats with orbital change capabilities are difficult to intercept. On militarization of space, NATO has made clear that although space was declared an operational domain, it has no intention to put weapons in space while upholding international law.

Notes

(i) Was a class of reusable cargo spacecraft developed by SpaceX.
(ii) Commercial Orbital Transportation Services.
(iii) Michael Collins passed away on April 28, 2021, at age 90.
(iv) Orbital Debris Quarterly (2021, 2) is published by NASA Debris Program Office (ODPO) and provides reliable information on the debris in outer space.
(v) Zarya.info is a website managed from the South Tyneside area in the UK.
(vi) In the ISS: Pyotr Dubrov (Roscosmos), Akihiko Hoshide (JAXA), Shane Kimbrough (NASA), Megan McArthur (NASA), Oleg Novitskiy (Roscosmos), Thomas Pesquet (ESA), Mark Vande Hei (NASA). These astronauts and cosmonauts are the crew of the Expedition 65 mission aboard the ISS. In the Tiangong: Space Station: Nie Haisheng (CNSA), Liu Boming (CNSA), y Tang Hongbo (CNSA).
(vii) As of 17 September 2021, over 88% of all satellites, probes, landers, crewed spacecraft, and space station flight elements launched into Earth orbit or beyond have been registered with the Secretary-General.
(viii) A geologic trainer for other Apollo moon-bound astronauts, the first scientist in a crew to the Moon.
(ix) The crew: NASA astronauts Mike Hopkins, Victor Glover, Shannon Walker, and Japanese Aerospace Exploration Agency astronaut Soichi Noguchi.
(x) Business Insider (BI) is an American financial and business news website founded in 2007.
(xi) See The Privatization of Space by Carolina Beirne and Sharanya Swaminathan. 6/24/2020.
(xii) Pilots Dave Mackay and Michael Masucci flew the spaceship. The crew was Virgin Galactic staff: Beth Moses, chief astronaut instructor; Colin Bennet, lead operations engineer; and Sirisha Bandla, vice president of government affairs and research operations.
(xiii) The Space Council is a joint meeting of the EU Council (Competitiveness - Space configuration) and of the ESA Council at the Ministerial level.

References

CIS. (2020). *Space threat assessment report 2020*. https://www.csis.org/analysis/space-threat-assessment-2020

Claeys, C. (2019, September 15). *Euroconsult's world satellite business week*. https://spacenews.com/eu-agency-starts-space-sustainability-initiative/

EU. (1975). *Convention establishing a European Space Agency of 30 May 1975*. https://www.esa.int/About_Us/Law_at_ESA/ESA_Convention

EU. (2016). *Joint statement on shared vision and goals for the future Europe in space by the European Union and the European Space Agency space strategy*. https://www.esa.int/About_Us/Corporate_news/Joint_statement_on_shared_vision_and_goals_for_the_future_ofEurope_in_space_by_the_EU_and_ESA

EU. (2019). *Space as an enabler*. Conclusions of EU Council, 28 May 2019. https://www.consilium.europa.eu/en/meetings/compet/2019/05/27-28/
EU. (2021). *Regulation of the European Parliament and the Council establishing the EU space program*. https://www.consilium.europa.eu/media/37659/st15767-en18.pdf
Fiott, D. (2021). *Securing the heavens*, 15 April 2021. https://www.iss.europa.eu/content/securing-heavens
NASA. (2018). *NASA's exploration campaign: Back to the moon and on to mars*. https://www.nasa.gov/feature/nasas-exploration-campaign-back-to-the-moon-and-on-to-mars
NASA. (2021). *NASA-axiom space agreement*. https://www.cbsnews.com/news/civilian-commercial-flight-to-space-station-moves-forward-with-nasa
NATO. (2018). *NATO Brussels summit declaration*, 11 July 2018. https://www.nato.int/cps/en/natohq/official_texts_156624.htm
NATO. (2019a). *PO (2019)0279 (INV), NATO overarching space policy*. https://www.nato.int/cps/en/natolive/news_167181.htm
NATO. (2019b). *NATO London declaration*, 4 December 2019. https://www.nato.int/cps/en/natohq/official_texts_171584.htm
NATO. (2021). *NATO Brussels summit communiqué*, 14 June 2021. https://www.nato.int/cps/en/natohq/events_184241.htm
Orbital Debris Quarterly. (2021, 2). *Orbital Debris Quarterly*, Vol. 25, Issue 2, June 2021. https://orbitaldebris.jsc.nasa.gov/quarterly-news/pdfs/odqnv25i2.pdf
PTBT. (1963). *Treaty banning nuclear weapon tests in the atmosphere, in outer space and under water (Partial Test Ban Treaty, PTBT)* signed on August 5, 1963. https://www.state.gov/limited-test-ban-treaty
Stoltenberg. J. (2019). Press conference after the Ministers meeting on 20 November 2019. https://drjustinimelsr.com/2019/11/21/press-conference-by-nato-secretary-general-november-20-2019/
Stoltenberg. J. (2020). Press conference Ministers meeting on 22 October 2020. https://www.nato.int/cps/en/natohq/opinions_178946.htm
UN. (1945). *Charter of the UN*, signed on 26 June 1945. https://www.un.org/en/about-us/un-charter
UN. (1967). *Outer space treaty. Treaty on principles governing the activities of states in the exploration and use of outer space, including the moon and other celestial bodies*. https://outerspacetreaty.org/
UN. (1979). *The Moon Agreement of 1979*. https://www.unoosa.org/.../spacelaw/treaties/intromoon-agreement.html
UN COUPOS. (1959). *Committee on the Peaceful Uses of Outer Space (COPUOS)*. https://www.unoosa.org/oosa/en/ourwork/copuos/index.html
UNGA. (1958). *UN General Assembly Resolution (d) 1348 (XIII)*. http://www.unoosa.org/oosa/oosadoc/data/resolutions/1958/general_assembly_13th_session/res_1348
UNGA. (1963). *UN General Assembly Resolution 1884 (XVIII)* adopted on October 17, 1963. www.unoosa.org/pdf/publications/ST_SPACE_061Rev01E.pdf
Woods, T. (2020). *Visualizing all of Earth's satellites: Who owns our orbit?* https://www.visualcapitalist.com/visualizing-all-of-earths-satellites/

A New Arms Race in Space? Options for Arms Control in Outer Space

Götz Neuneck

Abstract Modern societies are increasingly dependent on space-based services, making the preservation of the peaceful use of the space environment and the protection of critical infrastructure central to peace and security on Earth. New actors, programs, and technologies challenge current arrangements. The increasing weaponization of space, given the vulnerability of satellites, is a serious problem that can only be limited in the longer term by new rules, more transparency, confidence building and standards. The 1967 Outer Space Treaty provides the foundation for future space regulations, but it needs additions through confidence building, arms control and a better implementation of long-accepted standards.

Keywords Outer space · Missile defence · Anti-satellite weapons

1 Introduction

Near-Earth space is becoming increasingly important for peace and security policy on Earth. There is often the talk of a "New Space Age." Several trends are determining future space security: In addition to the classic space actors, i.e., the dominant USA, challenging Russia and ascending China as well as Europe, Japan and India, commercial companies are emerging which propagate easier and cheaper access to space, new satellite services in the communications and satellite sector or the exploitation of raw materials on other celestial bodies. Smaller countries are increasingly active in this area as access to space has become easier. Technological advancements, as ever, also can enable peaceful as well as military options. Future investments are always lured by high-tech fantasies and the adventure of "outer space," but also by calls for the using the space environment for military purposes. Space affaires are associated with international prestige, technical independence and technological leadership, as well as the ability to project power ("power through space"). Against

G. Neuneck (✉)
Institute for Peace and Security Policy, University of Hamburg, Beim Schlump 83, 20144 Hamburg, Germany
e-mail: neuneck@ifsh.de

J. Cayón Peña (ed.), *Security and Defence: Ethical and Legal Challenges in the Face of Current Conflicts*, Advanced Sciences and Technologies for Security Applications,
https://doi.org/10.1007/978-3-030-95939-5_2

the background of developing power rivalries between the USA, Russia and China, the competition in and around space is intensifying. In some fields of technology, especially those with a military background, arms racing can be observed. The mere assumption that the potential adversary is investing in key technologies is enough for enormous investments in military-relevant space programs such as missile defence, lasers, or Artificial Intelligence. The leading space powers also accuse each other of weaponizing space while openly testing anti-satellite capabilities. Approximately 20–25% of all satellites today are used for military purposes, i.e., reconnaissance, early warning, communications, missile navigation, etc. They are central to the coordination of global military operations and to weapons deployment (drones, cruise missiles).

Multilateral efforts to develop sustainable space norms for responsible behaviour by spacefaring nations are overdue. In the 1967 Outer Space Treaty, space is described as a sovereign common space whose "peaceful use" is to be in the interest of all states and "mankind as a whole". Since the beginning of space travel, there have always been international cooperation in addition to national and commercial interests due to the high costs and risks in this largely hostile medium: Emblematic here is the operation of the International Space Station (ISS) or joint planetary missions. This internationalist approach is in danger of fading away. More national security in space and protection of critical infrastructures is the order of the day.[i] The establishment of a U.S. "Space Force" (since 2019), increased Russian-Chinese cooperation, and the testing of "counter-space activities" by the three leading space powers are evidence of the preparation of a weaponization of space that can be used in the event of war or even trigger a war.

2 Legacy of the Cold War: Dual-Use, Restraint and Space Debris

The launch of the first Sputnik in October 1957 marked the beginning of the space age. Since then, near-Earth space has been accessible, and almost all nations now use satellites that fly over large areas of the Earth at different altitudes. The number of man-made objects has multiplied in the last decade. Currently, there are almost 7600 active and defunct satellites in space, which also enable commercial services: communication (TV, telephone, Internet, etc.), navigation (GPS, Galileo, etc.), etc. or Earth observation (weather, environmental monitoring, etc.). Today, by means of powerful launchers, many small satellites can be launched with one launch of a heavy space launch vehicle (SLV). This reduces the launch costs. Private companies such as Space X, One Web or Amazon are planning "mega constellations" (Starlink 12000 satellites) that will provide Internet services also to remote regions of the planet. It is estimated that in the next decade as many as 100.000 satellites will be added. This growth of orbiting satellites and the rising use by countries alone also increases the risk of collisions and electromagnetic interferences. Satellites are fragile structures that are quite easily affected by meteorites, kinetic objects and cyberattacks. Prolonged disruption or even failure of satellites could lead to serious problems on Earth if critical services are disrupted. But as their numbers increase, so does the problem of space debris. Space debris orbiting the Earth uncontrollably in various orbits is the result of rocket launches (burned out rocket stages) or broken satellites.[1] Because of their breakneck speed and small size, they pose a danger to active satellites, especially since they remain in orbit for a long time, depending on the orbit. The ISS has had to change its trajectory several times in view of the danger posed by meteorites or man-made space debris. In September 2009, two communications satellites, Cosmos 2251, and Iridium 33, collided at an altitude of 800 km, resulting in about 100,000 pieces of debris, of which about 2.200 are now catalogued. As the atmosphere thins due to climate change, less space debris is falling back to Earth. Thus, by 2100, the number of space debris could increase by a factor of 50 and pose a serious threat to the space environment.

2.1 *What Are Space Weapons?*

Already during the Cold War, the US and the Soviet Union have repeatedly pursued weapons programs and testing space vehicles with anti-satellite (ASAT) functions, but have not officially deployed any space weapons so far permanently. On the one hand, space weapons would be objects that are in space or can act into space and are designed to damage, render inoperable, or even destroy satellites.[ii] On the other hand, there are increasing developments of new manoeuvrable missiles (hyper-glide vehicles) that can launch on Earth, traverse space, and engage Earth targets again, i.e., a weapon that operates "from space". A prerequisite for the launch and operation of space weapons is both interceptors in outer space and the presence of ground-based infrastructure and associated data transmission and trajectory tracking. Table 1 provides an overview of the essential technological requirements for space weapons.

In principle, different technologies exist to hit, disrupt or render inoperable objects in space. The spectrum ranges from nuclear explosions in space to electromagnetic interference and kinetic (i.e., explosion or collision) weapons. A nuclear explosion initiated in orbit can damage or destroy satellites over a large radius due to the radiation released and pose a significant long-term problem.[iii] The use of "direct energy weapons" (such as lasers or microwaves, jamming), for example by blinding sensor systems, requires considerable technological experience.[iv]

Protection against such effects is also possible, but it adds weight to a satellite. Unilateral is hardening by shields or quick substitution of satellites or redundancy are possible concepts, but are expensive. An important category are kinetic effects,

[1] Space debris is divided into three categories: Particle diameters less than 1 mm, between 1 mm and 10 cm, and fragments greater than 10 cm in diameter, which are readily observable and mostly catalogued.

Table 1 Requirements for potential space armament

Access to space	Precursor technologies for space weapons	Weapon relevant experiences
– Launchers and launch facilities – Orbit and manoeuvring technologies – Satellite construction and mission management – Orbit tracking; tracking, telemetry and control/command	– Precision manoeuvres in orbit – Construction of small satellites – Power generation in space – Re-entry technologies – Global orbit tracking – Autonomous orbit tracking, – Communication and Control – Launch-on-demand – Optics and navigation systems	– Nuclear explosions (in space) – Interception of satellites and rockets – Laser experiments – Microsatellites – Cyber attacks – Jamming and electronic interference – Manipulation tools

which can easily be achieved in space by collisions due to the high intrinsic velocities of satellites. Anti-satellite interceptor missiles can be launched from Earth or stationed in space for longer periods of time and used in the event of war. Ground-launched missiles can also be used to directly target specific satellites as part of Ballistic Missile Defence (BMD). Satellites can also be rendered inoperable directly by external manipulation or an explosive charge. Another option is the increasing number of small satellites, which have limited manoeuvrability but can act like a "space mine." Larger "combat satellites" must be brought to the target through co-orbital manoeuvres, requiring fuel and time. These types require extensive testing, globally distributed ground stations, and years of technology development.

A final possibility is to disrupt or cut the data link to a satellite. Due to the high dual-use potential of space technologies, the leading space nations in particular possess such capabilities. The orbits of target satellites must also be measured, which requires radar or optical orbit tracking systems, i.e., extensive space surveillance. Satellites return cyclically, so they are easy to track. Due to their lightweight structure, they are very vulnerable and especially destructible at low orbital altitudes (LEO, Low Earth Orbit), also and especially by the ground-based missile defence (BMD) systems of the USA, Russia, China, and India, which are currently being developed.

2.2 *Military Use in Outer Space*

During the Cold War, military use of space was a key driver of superpower space programs. Many developments, such as in launch vehicles, earth observation, and satellite technology, illustrate the dual-use nature of space technology, as its operations can be used for military as well as commercial or peaceful purposes. As William

Burrows put it in its seminal "Story of the First Space Age": "Yet space could be reached only by rockets and the shame of it was that rockets were conceived time and again, expressly to wreak unparalleled destruction and kill large numbers of the very people whose enlightenment and salvation they promised"[v]. As early as the late 1950s, the U.S. and the USSR began also developing and testing kinetic anti-satellite (ASAT) technologies (see the timelines in Table 2). Nuclear tests in space (e.g., Starfish Prime 1962) showed that due to the nuclear triggered EMP effect satellites and communication services would be massively disturbed.[2]

From Earth, missiles can also kinetically destroy satellites through direct ascent attacks as part of missile defence interceptors.[vi] Another method is to put satellites into orbit, bring them close to enemy satellites, and then destroy the targeted satellite, a method that is slower and more predictable, but also achieves higher orbits. Such "orbital ASAT tests" were conducted notably by the USSR in the 1970s (see Table 3). With Reagan's SDI speech in 1983 and an assumed Soviet ASAT threat, U.S. ASAT developments intensified. Again, and again, military developments were pushed forward by both superpowers, such as armed space stations (USA: MOL and USSR: Almaz) or laser weapons. However, after some development and testing they were not deployed permanently at the end: too expensive to maintain, inefficient and too dangerous in case of crisis were the main arguments. Other military based technologies such as reconnaissance and early warning from space had a more de-escalating effect. They lowered the risk of misperceptions and of unwanted escalation in a crisis. This legacy is at stake today.

3 New Anti-satellite Tests: Counterspace, Electronic Warfare and Missile Defence

The international wake-up call for a new push in space weapons was the Chinese satellite test in January 2007, when the PRC succeeded in using a ground-based interceptor missile to destroy its own Fengyun-1C weather satellite.[vii] Subsequently in 2008, the U.S. used its SM-3 ship-based interceptor missile to strike a defunct U.S. satellite, demonstrating to China and Russia its already operational capabilities.

The development of "counterspace technologies" (ASAT, lasers, electronic warfare, etc.) has intensified since the late 2000s.[viii] Thus, satellite experiments for rendezvous purposes by the U.S., China, and Russia have also been increasingly observed. These "Rendezvous and Proximity Operations" (RPO) can be well camouflaged, as they can have civilian "service" purposes, such as refuelling or repairing other satellites. While it was recently believed that satellites in high geostationary orbits (36,000 km) were inaccessible and therefore safe, there is increasing evidence that all space powers are also testing space weapons for high orbits. While

[2] An "Electromagnetic Pulse" (EMP) triggered by a nuclear explosion in the vacuum of space produces electrical malfunctions in electronic devices through a broadband spherical wave in the absence of hardening.

Table 2 U.S. and Soviet/Russian ASAT and BMD programmes (selection)

Land	Years	Programme	Description	Status
USA	Late 50s, early 60s	„Bold Orion"; Project WS-199A	The ALBM on 19 Oct 1959 missed Explorer 6 by about 6.4 km	Air Force test of ALBM air-launched missile
USA	1957–1960	Project 505 „Mudflap": nuclear-tipped Nike-Zeus ground-air-missile as ASAT	8 missile tests until 1966	1967 terminated
UdSSR	Late 1960s to early 1980s	Co-orbital ASAT-programme «Istrebitel Sputnik»	First tests in 1968, resumption of tests in 1976 and 1977 with 4 tests each year	Officially out of service since 1993
USA	1960s–1975	Project 437: nuclear-tipped Thor-Missile-System	16 tests (1964–70), stationed at Johnston Atoll in 1964	Operation until March 6, 1979
USA	1983–1984	Homing overlay experiment	4 tests, one claimed as successful	First hit-to-kill tests against space objects
USA	1980s–1988	Project ASM-135 ALMV Missile onboard of a F-15, MHV	9 of 20 flight tests successful ($1.6 billion); Sole intercept of Solwind satellite on 9/13/1985	Air-based hit-to-kill-technology
USA	From 1999	Ground-based strategic missile defense (former National Missile Defense)	11 of the 20 hit-to-kill intercept tests have been claimed as succeeded	44 ground-based interceptors operational since in Alaska and Vandenberg
USA	From 2003	DART and XSS-11/12 (USAF) 2005; Orbital Express	Demonstration for autonomous rendezvous technology	Rendezvous and proximity technologies
USA	From 2006	Aegis System; SM-3	Interception of the inoperative satellite USA-193	14.2. 2008 test, regional operative
USA	From 1999	X-37B (US Air Force)	Secret test flights of prototypes 2010, 2011, 2012, 2015	In flight testing

(continued)

Table 2 (continued)

Land	Years	Programme	Description	Status
Russia	Strategic missile defense	PL—19 Nudol	Test Nov. 18, 2015, May 2016, December 2016, March 26, 2018, and Dec. 23, 2018	Direct ascent intercept

Table 3 Newer ASAT-testing by the US, China, India (See Notes[XX])

Country	Date	Target satellite	Interceptor missile	Collision altitude (km)	Number of debris	Estimated life time of debris
China	11.1.2007	FY-1C	SC-19	800	3.000	Several decades
US	20.2.2008	USA 193	SM-3	240	174	18 month
India	27.3.2019	Microsat R	PDV-Mk II	280	Appr. 300–400	Weeks/Month

the U.S. has been testing unmanned rendezvous technologies since 2003, China tested for example the Aalong-1 satellite in 2016, which can collect space debris with a robotic arm. Russia used the Luch satellite to approach several satellites parked in geo-stationary orbit. In 2020, U.S. Space Command accused Russia of releasing a Cosmos 2543 subsatellite to spy on a U.S. spy satellite. In July 2020, Russia was charged that this satellite ejected a projectile and used it to conduct an ASAT test. The U.S. accuses Russia and China of advocating arms control in space on the one hand, but secretly testing ASAT weapons on the other. The use of lasers for blinding satellites and electronic jamming against foreign satellites is progressing in these three space powers. They have been actively used in military operations. India has also entered the counter-space development. On March 27, 2019, the Prime Minister of India announced India's first successful ASAT ("Mission Shakti") test. In this operation, an ASAT interceptor had destroyed India's Microsat R test satellite about three minutes after it was launched by a ground-based rocket at an altitude of 300 km. An Indian spokesman claimed that the test was not directed against any nation and that the capability was acquired for deterrence purposes. Russia is testing the "Nudol" and "Contact" ground- and air-based interceptors for low-orbit interception. China conducted its own interceptor tests between 2010 and 2018. NATO's missile defence capabilities, which were first deployed by the U.S., thus also provide a regional ASAT capability and pose a threat to low-orbit satellites in the event of an armed conflict.

4 Threats and Space Doctrines: New Fighting Domains

The increasing importance of the space environment for military purposes is also underscored by relevant documents, programs, and the establishment of space command and centres of the leading space powers. In June 2018, then-President Trump declared, "We must have American dominance in space." The new 6th Armed Forces "Space Force" established by Trump proclaimed, "Humanity has changed, and the actions of our potential adversaries have significantly increased the likelihood of warfare in space".[ix] Outer Space and related to that, cyberspace is listed as new "warfighting domains" in the "National Space Strategy" released in March 2018, justifying new military space developments, and raising the prospect of a "thoughtful response".[x] In 2020, the Pentagon has developed its own "Defence Space Strategy" for maintaining "superiority in space". Russia and China are blamed for expanding the "weaponization of space" and their programs and doctrines are seen as coming strategic threats. At the same time, "space warfighting operations" are to be integrated into U.S. operational command. The Defence Intelligence Agency, in its 2019 Report "Challenges to Security in Space," also lists Iran and North Korea as challengers to U.S. superiority in space.[xi] At its last summit communiqué, NATO devoted a separate section to space.[xii] On the one hand, it wants to strive for responsible behaviour in space; on the other hand, an attack in space can now also be considered as an alliance case. A NATO Space Centre is under construction in Ramstein, which is intended to co-ordinate the space activities of NATO members. Yet, there are to be no NATO operations in space. Russia is continuing its old space tradition and has been developing various programs since 2010. Often, the purposes of various developments are unclear. Since 2015 Russia is also reorganizing its air and space defence. The Russian and Chinese space industries are largely state-controlled. Cooperation between Russia and China in both the civilian and military sectors is progressing. Beijing has decided that China wants to become a space power "in all respects." The space program is part of President Xi's "China Dream," encompassing three launches centres and operating its own space station. China's participation in the ISS was rejected long ago. Lunar and Mars missions underscore China's ambitioned goals in space. In July 2019, China published its first defence white paper since 2015, in which "space, electromagnetic space, and cyberspace" are considered a domain for national defence. Also, since 2015, these domains have been managed by the "Strategic Support Forces." While no direct ASAT tests by China have been observed since 2018, more non-kinetic tests (lasers, electronic attacks) appear to be taking place. There is no transparency between these space powers in the military sphere. Mutual suspicions and accusations dominate statements and doctrines. The unmanned Mini-Space Shuttle X-37B has already carried out six long-term missions with unknown purpose and worries China and Russia as much as vice versa the ASAT tests of the PRC and Russia. Hopefully, the Biden administration will make serious efforts to advance international regulations, establish new rules for space, and provide greater transparency and predictability for military programs.[xiii]

5 International Rules and Arms Control in the "Interest of All States"?

The Outer Space Treaty of 1967 (Outer Space Treaty) is the first binding multilateral treaty and thus the Magna Charta of international space law, extending the UN Charter to outer space. In the preamble, it emphasizes the "common interest of all mankind in the progressive exploration and use of outer space for peaceful purposes" and declares it a "matter for all mankind". Accordingly, outer space is a sovereign common space, the use of which must be "in the interest of all states" and of "mankind as a whole". Accordingly, security in outer space cannot be pursued exclusively in the national interest of one state. Article IV explicitly prohibits the deployment of weapons of mass destruction in space, but not the stationing or use of "conventional weapons" or the electronic interference with satellites. An additional protocol that also prohibits the use of satellites as destructive weapons would certainly have a preventive prohibition function for many states. Nevertheless, given the diversity of ASAT weapons, a definition of a ban on space weapons has not yet been achieved. Other important international treaties were negotiated during the "Golden Age" of space law between 1967 and 1984 (see Table 4). For example, the 1976 UN Registration Agreement requires states parties to report information about the launch date, location, and purpose of the mission to a UN agency in Vienna. Unfortunately, the reported data from the world's space powers today is inadequate and reflects the intentional lack of transparency of military relevant space missions. This behaviour must

Table 4 Major space agreements

Treaty	In force since	Ratifications/signatures	Subject
Outer Space Treaty	10.10.1967	104/25	Magna Carta of Space Law, which permits the exploration objects and use of outer space only for the benefit and in the interest of all states
Space Rescue Agreement	3.12.1968	94/24	Commitment to rescue astronauts after a disaster
Space Liability Convention	1.9.1972	92/21	Liability requirements for damage caused by space object
Space Registration Agreement	15.9.1976	62/4	Obligation of the contracting states to register rocket launches, orbital parameters and payload
Moon Treaty	11.6. 1984	16/4	Prohibits military use of the moon and other celestial bodies

be significantly improved. Other arms control agreements (prohibition of nuclear tests in space or the use of satellites for verification purposes) include space, but are visibly eroding. The termination of the ABM Treaty in 2001 by the USA, allow that strategic missile defence can be used for anti-satellite actions in low orbits in case of war or emerging armed conflict. Crucial for this are not only earth-based interceptors but also a functioning space surveillance by radar and optical methods, which are only in the possession of a few space powers. Space is a transparent medium and, in principle, it is not so complicated to monitor and to verify orbiting satellites from earth and space. Most of these capabilities are in the hands of the leading space powers.

Proposals for a comprehensive prohibition regime for space weapons were developed in the 1980s to keep outer space weapon-free on a multilateral basis. They range from unilateral commitments by states to informal agreements, confidence- and security-building measures (CSBMs), and mandatory codes of conduct to comprehensive prohibition regimes.[xiv] In several international fora, such as the Geneva Conference on Disarmament (CD), the issue of Prevention of Arms Race in Outer Space (PAROS) has been on the agenda since 1982 without producing a concrete result. In 2008 and updated in 2014, Russia and China had proposed a treaty, "Prevention of the Placement of Weapons in Outer Space" (PPWT)[xv] which commits states parties to "not place weapons in outer space." Criticisms focused on the inadequate definition of a space weapons, the difficulties of verification, and the exclusion of earth-based ASAT weapons or non-disruptive techniques. Even if the arguments are justified, hardly any serious proposals have been presented by the Western side to overcome these hurdles.

In the United Nations, resolutions on the prevention of the space arms race have been adopted annually since 1981, emphasizing the need and desire of the international community to prevent the weaponization and a dangerous arms race in space.[xvi] At the UN level, a UN Group of Governmental Experts (GGE) was established, which developed voluntary transparency and confidence-building measures for space activities in its 2013 report.[xvii] In December 2017, the UN General Assembly established another UN GGE to develop "elements for a legally binding instrument to prevent an arms race in space." Despite a report rich in content, there was no consensus for its adoption. While the negotiating states agree that there is a significant need for rules, there is currently no comprehensive regulation in sight (Table 5).

The International Code of Conduct (ICOC) of the European Union was initiated in 2008 as a non-legally binding proposal. This focused on the avoidance of space debris and collisions in space and wanted to promote the "responsible behaviour of states in space". Australia, Japan, Canada supported the proposal, but not the U.S., Russia and China, as well as many Latin American and African states, as it remained too non-binding and vague. The ICOC failed at a UN conference in New York in 2015. Since then, the EU has focused on elements of a future Space Traffic Management (STM) regime, e.g., in the framework of the EU-ESA Space Council. The EU could now put forward a new proposal to ban ASAT tests and establish a new instrument in space policy as an initiator and facilitator, including CSBMs and verification mechanisms.[xix]

Table 5 New initiatives for the improvement of space security[xviii]

	PPWT	ICOC	LTS guidelines	TCBM (UN)
Discussion since	2008	2009–2015	2010	2011–2013
Forum	Conference on disarmament	On invitation of the EU	UN COPUOS STSC	UN GGE
Initiators	Russia/China	European Union	UN COPUOS STSC	Russia/UN GA
Subjects	Ban on space weapon deployment	Avoidance of space debris and collisions in space	Guidelines for new actors in space	Development of TCBM
Rejected by	USA and Western States	Russia/China, NAM, Africa, Latin America	–	In discussion

Another forum is the civilian-oriented UN Committee on the Peaceful Uses of Outer Space (UN-COPUOS), in which some 90 member states voluntarily developed 21 "guidelines for the long-term sustainability of space activities" from 2010 to 2018.[xx] Such "best practices" can help avoid collisions with space debris, increase awareness of hazards in hazardous situations, or just share official data. Involving the booming private industry is essential here, because protecting their expensive space assets is of great economic importance to operators. However, military aspects are kept out of UNCOPUOS, so a solution to the ASAT problem seems not to be possible here.

Under the leadership of Great Britain and some like-minded countries, including Germany, a new UN resolution 75/36 "Reducing space threats through norms, rules and principles of responsible behaviours in outer space" was introduced in the UN General Assembly (GA) in December 2020. States submit national reports on the subject here, which the UN Secretary-General is to summarize in a report. The international trend is to move away from object-based to behaviour-based prohibitions. There are good reasons for this, but in the end, binding commitments that verifiably prohibit space actors from attacking specific objects will also be necessary. There is hope, however, that governments will realize that major disasters in space or acts of war in space that have major impacts on Earth itself are not in the longer-term interests of the world community.

The Trump administration has consistently stated that it does not want to agree to treaty arrangements that threaten its supremacy. A Trump administration directive on cyber principles for space systems advocates "forward defence" and requested in the U.S. Congress the development of counter-space weapons that would cost hundreds of millions of additional dollars. New U.S. Defence Secretary Lloyd J. Austin, at his 2020 confirmation hearing, advocated more spending on space platforms, embracing commercial sector innovation, and maintaining the U.S. technological edge against the backdrop of "pacing China."

While arms control in space must be based on the principles of international space law, additional bilateral arms control arrangements among the world's leading space powers can help prevent acts of war in space and reduce the threat to space objects.[xxi] The hope was in the context of preventive arms control through preemptive bans on certain weapons developments from being developed and tested in the first place (See Notes[xiv]). However, current developments in the field of laser weapons, electronic warfare and ASAT technologies stand in the way of this and are forcing the progressive weaponization of space. All space powers proclaim that an arms race in space must be avoided, but serious steps to establish sustainable and lasting rules in space are hardly visible so far.

In July 2020, the U.S. and Russia held their first bilateral round of talks on "regulating the militarization of space" in Vienna since 2013. After extending New START for 5 years earlier this year, Presidents Biden and Putin agreed to a "robust and well-balanced" Strategic Stability Dialogue at their June 2021 meeting. This is intended to lay the groundwork for future arms control and risk reduction measures between the two superpowers. The agenda includes further nuclear disarmament of the high warhead ceilings as well as the space and missile defence issues. Confidence building measures to prevent a nuclear war which can be triggered by an attack on critical infrastructures in space and a prohibition of deploying space weapons to interfere with National Technical Means in space would be a big step forward. Space-based early warning components for missile attacks or earth observation and radar satellites for verification are key space assets that must not be attacked even in the event of a crisis. The creation of "keep out" zones or a non-attack obligation for strategically important satellites would be a first stabilizing step. A legally binding agreement to ban ASAT tests would be a major step forward. In their statement, Biden and Putin underscored the Gorbachev-Reagan formulation that "a nuclear war must not be won and never fought." Similarly, a war in space would undermine international space security significantly and create much new space debris. The leading nations in space, led by the U.S., Russia, and China, must move forward here with binding, risk-reducing measures and strengthen security for all objects in space and the future of humankind.

Notes

(i) Harrison et al. (2018).

(ii) Neuneck and Rothkirch (2003).

(iii) See f.e. Wright, Grego and Gronlund (2005).

(iv) Burrows (1998) p. 4.

(v) Stupl and Neuneck (2005).

(vi) On the combination of missile defence and space defence, see Neuneck et al. (2015). p. 135–153

(vii) Neuneck (2008).

(viii) Secure World Foundation (2020).

(ix) US Space Force 2020: Space Power Doctrine for Space Forces, US Space Force Headquarters, June 2020 https://www.spaceforce.mil/Portals/1/Space%20Capstone%20Publication_10%20Aug%202020.pdf.

(x) U.S. Department of Defense 2018, National Defense Strategy, Washington D.C., p. 3/6.
(xi) Defence Intelligence Agency, „Challenges to Security in Space", January 2019 https://www.dia.mil/Portals/27/Documents/News/Military%20Power%20Publications/Space_Threat_V14_020119_sm.pdf.
(xii) Brussels Summit Communiqué issued by the Heads of State and Government participating in the meeting of the North Atlantic Council in Brussels 14 June 2021, Nr. 33. https://www.nato.int/cps/en/natohq/news_185000.htm?selectedLocale=en.
(xiii) Samson and Weeden (2020).
(xiv) See Neuneck and Rothkirch (2006b).
(xv) PPWT stands for Prevention of the Placement of Weapons in Outer Space and of the Threat of Use of Force against Outer Space Objects. See details under: https://undocs.org/pdf?symbol=en/CD/1839.
(xvi) Meyer (2020).
(xvii) Examples here include the publication of national space policy, risk reduction notifications, and reciprocal launch site visits, etc.
(xviii) Kim (2017).
(xix) Raju (2021).
(xx) Secure World Foundation (2019).
(xxi) Wolter (2005).

References

Burrows, W. (1998). *This new ocean. The story of the first space age.* Random House.

Harrison, T., Johnson, K., & Roberts, T. G. (2018, April). *Space threat assessment 2018.* Centre for Strategic and International Studies. https://www.jstor.org/stable/resrep22469

Kim, Y. (2017). *Analysing the recent multilateral discussions on outer space security.* Fact Sheet Hamburg IFSH. https://epub.sub.uni-hamburg.de/epub/volltexte/2020/101648/pdf/IFAR2_FactSheet10.pdf

Meyer, P. (2020). Arms control in outer space: A diplomatic alternative to star wars. In J. M. Ramirez & B. Bauza-Abril (Eds.), *Security, and the global commons.* Springer.

Moltz, J. C. (2008). *Asia's space race national motivations, regional rivalries, and international risks.* Columbia University Press.

Neuneck, G. (2008). *China's ASAT test—A warning shot or the beginning of an arms race in space? Yearbook on space policy 2006/2007: New impetus for Europe* (pp. 211–224). European Space Policy Institute. Springer.

Neuneck, G., Alwardt, C., & Gils, H. C. (2015). *Raketenabwehr in Europa.* Nomos.

Neuneck, G., & Rothkirch, A. (2003). Space as a new medium of warfare? Motivations, technology and consequences. In *Changing Threats to Global Security: Peace or Turmoil—Proceedings of the XV Amaldi Conference* (pp. 163–189), Helsinki, 25–27 September 2003. Finnish Institute for International Affairs.

Neuneck, G., & Rothkirch, A. (2006a). The possible weaponization of space and options for preventive arms control. *ZLW German Journal of Air and Space Law, 55*(4), 501–516.

Neuneck, G., & Rothkirch, A. (2006b). Weltraumbewaffnung und Optionen für präventive Rüstungskontrolle. Deutsche Stiftung Friedensforschung DSF Nr.6, Osnabrück 2006, p. 9.

Raju, N. (2021). *A proposal for a ban on destructive anti-satellite testing: A role for the European Union.* EU Non-Proliferation and Disarmament Papers. EU Non-Proliferation and Disarmament Consortium No. 74. https://www.sipri.org/publications/2021/eu-non-proliferation-and-disarmament-papers/proposal-ban-destructive-anti-satellite-testing-role-european-union
Samson, V., & Weeden, B. (2020). Enhancing space security: Time for legally binding measures. *Arms Control Today, 50*(10), 6–13.
Secure World Foundation. (2019). *The UN COPUOS guidelines for the longterm sustainability of outer space activities.* Factsheet, November 2019. https://swfound.org/news/all-news/2019/06/guidelines-on-long-term-sustainability-of-outer-space-adopted-by-un-copuos/
Secure World Foundation. (2020, April). B. Weeden & V. Samson (Eds.), *Global counterspace capabilities: An open source assessment.* https://swfound-preprod.azurewebsites.net/media/206957/swf_global_counterspace_april2020_es.pdf
Stupl, J., & Neuneck, G. (2005). High energy lasers: A sensible choice for future weapon systems? *Security Challenges, 1*(1), 135–153.
Wolter, D. (2005). *Common security in outer space and international law.* United Nations Institute for Disarmament Research (UNIDIR).
Wright, D., Grego, L., & Gronlund, G. (2005). *The physics of space security. A reference manual.* American Academy of Arts and Sciences.

Cyberspace, Artificial Intelligence, and the Domain of War. Ethical Challenges and the Guidelines Proposed by the Latin American Development Bank

Fernanda Navas-Camargo and Carlos Alberto Ardila Castro

Abstract The past decades have been characterized by significant technological developments, which have allowed both the generation of a revolution in military affairs and the emergence of new armed conflict scenarios. Until recently, conflicts between individuals' societies and states had developed in a fundamental dimension regarding the ends, means, and modes of adversaries linked to geographical environments. With the emergence of cyberspace, these scenarios have changed to virtuality generating challenges to the proper safeguard of Security and Defense. Exposure of sensitive personal information has become one of the most feared weapons. This chapter focuses on a general review of the challenges posed by the emergence of cyberspace scenarios of armed conflict, emphasizing the ethical challenges that emanate from it. These new scenarios have been developing under the model of "anything goes," and the existence of restrictions on the use of force and the damage caused to the adversary is practically nonexistent. The purpose of this research is to examine these challenges seeking to find opportunities for the construction of public policies in the area of security and defense to strengthen the legitimacy of the States and present the guidelines proposed by the Latin American Development Bank to make more efficient and ethical use of Data and Artificial Intelligence. Chapter presents the research project's results entitled: Evolutionary Trends of Security and Defense Policies in the Americas; assigned to the Logistics and Military Administration research line, carried out by the research group "Centro de Gravedad." The research was funded by the Escuela Superior de Guerra "General Rafael Reyes Prieto".

Keywords Cybersecurity · Sovereignty · Territory · National security · National defense

F. Navas-Camargo (✉) · C. A. Ardila Castro
ESDEGUE, Bogotá, Colombia
e-mail: johanna.navas@esdegue.edu.co

C. A. Ardila Castro
e-mail: carlos.ardila@esdegue.edu.co

J. Cayón Peña (ed.), *Security and Defence: Ethical and Legal Challenges in the Face of Current Conflicts*, Advanced Sciences and Technologies for Security Applications,
https://doi.org/10.1007/978-3-030-95939-5_3

1 Introduction

The development of new technologies and the reign of the Internet over human interactions has constantly been evolving, updating, and permeating all spheres of human life without any distinction of gender, nationality, age, or level of education. Along with the individuals, the States and the International System have overseen becoming the referents of guaranteeing the rights of all citizens of the planet by being the ones who have needed to implement the most transformations starting by the necessity of redirecting the basic concepts of *territory* and *sovereignty*, which have been blurred over time.

Cyberspace is the area to which all these modern changes have been associated. In it, calculation and reasonability are introduced as bases for work development. The essential thing is to build an artificial intelligence machine that responds to the given and functional objective under the defined means. The space where all the known advances have been driven and that continues to be in constant transformation is often called a space without sovereignty, where any interaction between human beings, their new ideas, and the materialization of them without any limit is possible, as Barlow highlights in his document called the "Declaration of Independence of Cyberspace" (1996).

This has triggered the movement of globalization, which has been recognized as an invisible phenomenon that influences the growth or decline of all spheres of life, such as politics, social and economic, among others. It is called the only system of connections presented through markets of goods, capital, and information flows, among others, which has grown exponentially because of the management of social networks and technological advances. All these topics of interest have been globally penetrated handled. The recognized force that globalization continues to maintain to this day corresponds to the management of information and how it can be exposed by addressing discourses of power that can help or affect either positively or negatively these spheres (Cooper, 2010).

These new forms have been used to maintain, prolong power, and even help to position it. But this concept per se has not changed and is still recognized as a strategy. As Foucault describes it, power is exercised by taking advantage of the resources you have. In this sense, the more devices or means an individual has, the easier it is to recognize who holds power and which ones are their interests. In the same way, the greater the chances of maintaining control, whoever has most of the technological means co-opted and functioning under their conditions.

Thus, globalization is intrinsically related to promoting greater transparency in the knowledge of the models of governments and social control, which translates into the development of the media and transmission of information, thus using new technologies that can be replicated and used directly by the society. The social debate becomes an active part of the issues discussed in the public agenda by acknowledging the discourses that have been established of a given subject, its incidence, and the new actors that become relevant (Campillo, 2007).

The process of modernization has come hand in hand with creating new scenarios and actors. How traditional canons such as politics, the economy, and society were recognized have changed along with this process. Power relations were influenced by identifying new formations of social structures that have been transformed by these new dynamics of time and space, thus elucidating new forms through which any type of knowledge, relationship, and understanding of social reality, among other factors, can be obtained. This process has been determined to recognize a new reality (Castells, 2009).

Society has welcomed the changing dynamics, which have brought with them the incorporation of facilities for each of the activities of the human being. In the same way, this new technological wave called globalization has given new scenarios wherein the life of the human being has been intertwined with the different digital networks. It has also responded to new challenges for States wherefrom its concept it was not prepared to regulate all the scenarios in which the human being develops, discuss and debates on rights, and new arguments continue to be presented for the construction of ideas, not only of a State but in general of the entire international system.

As a new reality, it must be studied from all branches of science with the interest of managing or controlling it without trying to generate resistance or confrontations in the face of a phenomenon that is impossible to stop and that, when contained excessively, can cause negative results. In this sense, the States, despite being on many occasions the ones that to a greater extent may feel affected by the social changes that this generates, must still respond to these dynamics by focusing on development, whereby not entering these, any country can be involved in its progress, mainly by exercises such as those that occur in the economy whereby not participating in the new forms of trade, positioning, competitiveness, and place within the system begin to be lost.

Other modifications that are unleashed as a result of the new cyber zones in the States respond to seek better management of these new mechanisms, since these new scenarios have not only allowed to establish relationships from anywhere in the globe but also give rise to different readings of the same phenomenon, to the possibility of knowing confidential information and even making visible new threats. Therefore, it has been necessary to regulate this new scenario so that none of the actions that take place in the network can transgress any right of the other. In the same way, this regulation is not easy to implement by considering many factors affecting its proclamation's fundamental rights such as free expression.

The expansion of technologies and interconnection has caused a dependence on them, which is maintained in civil and armed structures such as the public force according to the needs of the State. As a result of this, new designs begin to be implemented to recognize the new space in which disputes can be generated and where it is necessary to establish security, which has become a situation of interest throughout the international community; since the processes have changed to the point at which it is required to make a difference of the possible war scenarios before and after the existence of globalization.

2 Wars Across the Time

It is also necessary to emphasize that the attacks and political and strategic confrontations play notable roles in destabilizing a nation-state by using cyberspace. Cybercrimes represent a technification and modernization of the dynamics of War, but the evident mutation in several characteristic elements of "traditional" wars, which leads to addressing some aspects of belligerence that occurred in diverse spaces. These have risen to social and political connotations and the development of theories that allow a better understanding of the reality of the various societies.

First, one must detach from the concept of conflict a negative understanding of the term or directly related to War. In terms of Estanislao Zuleta (2015), War is due to the confrontation of two actors who oppose their interests. However, a notion of the inevitability of conflicts is presumed since human beings by nature have opposite interests, which leads to the development of opposing perspectives, which translate into inevitable confrontations that can eventually escalate into wars (Zuleta, 2015).

However, "traditional" conflicts involve characteristics that are essential in the understanding of phenomena such as the differentiation of wars between external and internal agents, the recognition of the territory in which hostility occurs, the handling of weapons, among others; in which it can be referred to the concept of "Civil War," which is particularly diffuse since within the different disciplines of the social sciences several interpretations of it are presented since it is used very regularly to refer to any internal situation that is of an armed nature but that constantly evades the understanding by the "Civil" component.

Therefore, for the Colombian case, the most coherent definition is that of Kalyvas, allowing to go beyond the traditional debates associated with ideological and political aspects, and defines it as an armed confrontation, between parties subject to a joint authority at the beginning of hostilities, within the borders of a sovereign political unit, and recognized. It is pertinent to note that this definition is agnostic in the face of the causes, motivations, and objectives of War (Kalyvas, 2008).

Following Vargas, in his text "Civil War in Colombia: the case of Barrancabermeja," it is crucial to add to this definition, citing Kaldor as a conceptual reinforcement; another distinctive feature to be able to correctly delimit the phenomenon: while in conventional wars the capture of the territory is achieved from its military occupation, in civil wars it is achieved from the political control of the population. In other words, in civil wars, it is possible to control a territory only when it is ensured to the majority and exclusive collaboration, voluntary or involuntary, of the population with an armed group (Vargas, 2009).

This allows us to outline elements that have been the basis of any confrontation, for example, private property, the possession of goods, and the dominion over the territory, which in terms of Sun Tzu (2010) (s, f) are elements that require special attention to dominate in the War, since it grants privileged positions and represents the experience of emotions of War that are put at the service of the present organizations.

Territoriality represents a cross element since the interstate wars of Europe constituted a primordial component for the construction of nation-states because the State is the privileged scale and unit of regulation and social emancipation (Sousa Santos, 2002). The historical configurations of these wars generated an international paradigm where coexistence and recognition among equally sovereign States, generating international law and new ways of preserving themselves in the global system, with the constant of "Peace" and "security."

Now, with social transformations and technological advances, the panorama has changed. The elements that determine excellent references such as Sun Tzu have vanished, and the new reality has eliminated borders, territorial spaces, firearms, and even the recognition of the counterpart. The reason? War has moved to areas that are not necessarily physical and in which it is not necessary to use ammunition to cause damage. Instead, they use telecommunications tools and all the dynamics of information flow related to globalization, which has brought a weapon that can cause damage more decisively.

This is the world in which information, the economy, and power relations are moved and regulated; the traditional aspects that served to understand the facts and therefore direct the strategies to weigh an attack and even respond are now gone.

The dynamics have been transformed to the point where an attack can be received from anywhere globally, so it is difficult to track or safeguard from any activity and even more complex to recognize the author of it. Nowadays, it is possible to act without leaving any trace or clues concerning the control of networks, a situation that also affects the proceeding of the justice of each State, being almost impossible to determine the authorship of a crime and therefore making it impossible to pursue, where he is, how to guarantee that it is his identity and not a decoy, finally, these are aspects that had not been taken into account for security and defense issues in each nation and that have now taken great relevance and positioning given the power to new agents who intend to serve as a shield from external attacks.

The new reality responds to the need of establishing within the security and defense policies from each of the nations the concept of cyberspace, which must be regulated under specific terms such as freedom and security, among others, thus ensuring the rights of all people who live in the international community and constantly develop relations in cyberspace, using the conditions and elements that are in place to recognize this new scenario and within the possibilities to face the new threats that occur in this environment.

It is almost impossible to ensure that the actions intended to be imposed concretely on each State and even jointly by the organizations that make up the international community; can help contain all the threats and actors that have begun to elucidate in the new landscape. The dynamics of War are continuously transformed with new technologies, as there is no constancy or at least security in the face of what is being faced, it is impossible to generate actions that can accommodate all these dynamics, even so within this reality, the concept of cyber security has been imposed. In the following paragraph, a revision of the idea is presented.

3 Cyber Security

Adapting to the new reality means greater digitalization, especially of information and management of resources, thus hoping to make the processes more efficient and effective. For a State, generating all its processes under a purely digital environment means facing an extreme degree of vulnerability, which promotes new scenarios of uncertainty and threats, which has led to considering the issue of cybersecurity as fundamental for States linked to the care of their interests. This implies that part of the nation's resources is directed to explore cyberspace in the search to generate knowledge that contributes to creating ways that serve as a shield under these new tools, infrastructures, and cyber activities without fear of an intrusion or breach of security against information. One must remember that much of the data is highly discretionary, classified, and involves critical data such as the handling of nuclear weapons and other issues that could trigger a war.

These acts do not involve only social or economic issues. Still, they have now become spaces prone to threat management, implying new transformations of War that it is essential to begin to address. There have already been cases where illegal and even terrorist groups use the Internet to organize and act in the form of a network, issues of the impact that are sometimes only believed to have a measure which response to the limitation and surveillance of the use of networks, a notion that although it can be used as an exercise in which it is intended to maintain control and power, generates a constant doubt about how these limitations that are given with justifications such as preserving privacy can affect notions such as free expression and decision-making power in a situation by having limited access to all the information needed about a given fact, and that recalls to figures such as Foucault's panopticon (Campillo, 2007).

In the same sense, Klaus Schwab (2017) recognizes that the fourth industrial revolution associated with all the management of technology generated a much more significant social gap since not everyone has the ability to acquire the tools that allow them to have information, continuously update themselves, and know the reality that surrounds them; as well as a change like threats to international security since new ways are presented so that individuals regardless of their distinctions, of age, race, color, and even nationality, can come to know more and more ways to harm the other on a large scale, thus producing a greater sense of vulnerability and uncertainty by not being clear about where the attack may come from and what type it is.

It is clear, then, that the use of the territory and sovereignty are two of the essential distinctions by which a State is recognized, while its power attributes to it acting under the making of rational decisions that allow it to defend its territory and its inhabitants, from any threat that it catalogs of going against its interests. The violation of security through the network has repercussions even for Human Rights, with the destruction of platforms, the violation of privacy, and the subsequent derivation of acts that allow under a cyber attack to denigrate the human being. It shall be asked, to what extent sovereignty acts and on which actions are possible to regulate an intangible space?

War conflicts have evolved rapidly with the advent of these new technologies. With the emergence of cyberspace, the confrontation of interests appears again in a field much more linked to development, then addressing the situation from a recognition of Cyberwar oriented to the conduct of military operations that are associated from their expertise and with the relevant information within their training, to recognize the information channels tending to destroy or control the communication systems of the adversary, with which data is obtained for security and basic actions are designed when analyzing future scenarios, to such a dimension that it is necessary to contemplate even when these may present large-scale conflicts between two or more societies.

The definitions of cyberspace are being assumed under in-depth investigations by agents of the State, for example, in Spain, the first official definition of cyberspace was given in February 2013 in the Ministerial Order, by which the Joint Command of Cyberdefense of the Armed Forces was created, describing the global and dynamic domain of the technology that mainly handles information, from the telecommunications networks and base systems used by each institution or agency, which live in constant anxiety as there is no guarantee of the care of the data.

Recognizing then the role that technological development has taken in the emergence of a space of relationship where the interaction of information and communication is easily generated without any barrier, where there is no room to think about others; the subject has taken so much boom that in the regulation on cybersecurity, this is represented as the activities necessary for the protection of networks and information system, addressing all actors that may be affected by cyber-threats, either in institutions or civilian members. Threats such as illegal access to personal computers, data theft, and card cloning are problems that have become common and difficult to address, which requires regulation that is in line with the need to face both small-scale actions as well as the most important large-scale actions such as intrusions to institutions like banks, ministries, among others, which can leave much more telling implications in society (Ardila Castro & Cubides Cárdenas, 2017; Ardila Castro, 2018).

The work of organizations to build this cybersecurity is arduous. Each of the nations tries to identify real solutions to new practices but has no limitations from the field of expertise. The area can be addressed from anywhere to generate damage. So, it is necessary to implement a cooperation mechanism that allows States to combat cybercrime, guaranteeing privacy rights to society, avoiding actions that could involve as many parts as possible, and breaking the social compound not only nationally but also internationally; the proposals as a whole must be confronted with the reality of cyberspace and assessed for relevance (European Commission, 2017).

The world in which we find ourselves has been so transformed that even international organizations have taken a stand on the new cybernetic context, especially the United Nations Group of GovernmentalExperts. The latter in 2013 recognized the application of international law, where the conception of cyberspace has been included in the Charter of the United Nations. This was an important milestone where digital peace and security were recognized. Therefore cyberspace is no longer a "lawless land," It can now be regulated by international law, identifying and interpreting existing rules in the question of their relevance and limitations in cyberspace (Pana, 2021).

So the recognized cybersecurity is the set of procedures, strategies, and elements that are imposed to protect the information that is generated through networks or systems that are translated as computers, servers, mobile devices, among other electronic devices; the cybersecurity approach has multiple protection sections, layer distributed in the networks or programs, with which it is intended to protect the information and the use of it under the creation of a unified system of effective defense against threats and cyberattacks.

4 Actors of Cyberspace

Access to the Internet has become a new Human Right, recognized worldwide, advocating its importance for freedom of expression, opinion, and exchange of information, without limitations. Still, in recent years, this digital universe has become a space of confrontation, not only between States but with the emergence of new non-state actors, which with their actions can destabilize any State or industry, which corresponds to the growing use of ICTs under harmful intentions by organized groups, Extremists and even terrorists, this phenomenon corresponds to a knowledge of new agents that have begun to have great relevance within the international panorama.

Among the first actors that are recognized are raised the state agents, these as members of the State have the objective of consolidating harmony and territorial peace, helping to impose actions that aim to maintain the political and social order against actual or supposed threats according to the interests arranged, these are usually recognized in agents such as the army and police, entities that under law and regulation have a monopoly on force and use it to maintain what is established in the national system and its national identity.

These state actors have been transformed because of the new conditions in which social relations are presented, such as cyberspace, where national identity begins to permeate and therefore be lost. In contrast, it has been necessary to generate new networks, mainly technical in the face of systems and technological advances understanding that these can cover and safeguard the relevant information, where state officials act and directly influence the political or economic life not only of the countries but of the entire international system (Zárate Botía, 2008).

That is why, within the institutions that correspond to state agents, agents who are only dedicated to the care, management, and protection of state networks, their security and protection against the information that is presented have begun to be professionalized, this is one of the strategies that have emerged in the face of the new cyber contexts, which have been implemented in search of National Security and Defense.

Also, non-state actors have obtained relevance within international relations that offer elements that allow establishing a determining role for agents such as NGOs that have appropriated issues like the environment, the protection of human rights, among others, which would enable under given conditions to infer in aspects or agenda items of the international system, especially in conflict issues these are usually associated with neutral agents that allow diplomacy between the parties and seek practical solutions to confrontations.

International non-state actors have the power even to generate actions of domestic social movements in contexts of state repression. They have also used modernization and networks as a source to add their legitimacy, exercise dynamics that correspond to their power, and thereby maintain their ideal; their primary interest is not related to the practices of States, since they properly have their ideals marked by which they work with the necessary tools; however, their approach and interest in human rights have made them in one way or other agents of interest in the foreign policies of States since this issue has been implemented in all internal policies (Bitar Giraldo, 2006).

Under this order of ideas, the state actor is formed as an indispensable agent mediator. With its interests that respond to the needs of the entire international society, seeking to safeguard the violations of the rights that may arise, therefore, with the use of these new technologies and the management of networks, the media power has attracted so much interest from the society that it became a determining role to the change in the behavior of such States, creating favorable conditions for the collective action of movements and the recognition of domestic social debates that influence security.

Given that, these new actors have tools that within this new system try to coordinate and generate security aspects for the care of the system and those that converge in it. Therefore, this new agent uses the new cyber network and relates its action with all the new actors presented and those that have always remained in the system as a stable agent with influence.

There are also parastatal actors. In their implicit definition, these correspond to mixed actors that can come to cooperate with the State but are not part of its administration, even so with time, its concept has changed, becoming more involved in illegal agents who use force excessively in search of an end of their own, which has been able to permeate and affect society in general. As Aguila and other authors acknowledge, this process has affected the system and created fragmentation in the social fabric (2016).

In this sense, far from being actors who seek good for the international community, some of them have become agents that have been destined as a threat to the system, affecting the stability of the States and the implementation of Human Rights throughout the globe, the parastatal groups became with social changes a danger to the community and with the use of new tools they have been able to increase their capacity, testifying their attacks and Permeating all spheres of human security (Waldmann, 1995).

Finally, along with the development, new unique agents are recognized under the professionalization of specific sciences, such as systems engineering, which can also be associated with a threat against States and the international system. These have become agents of utmost importance, such as hackers or specific groups, which with the use of their tools and sufficient knowledge can violate the security of the systems of nations, addressing private information or generating new attacks through the use of mass media implementing discourses that affect the State or international system and its interests.

Groups like Anonymous that have gone viral in the international community manage their notion and rationalization as well as their way of acting under the ideals they drive, associated with their level of justice management of the use of their resources, thus seeking a counter position to the impositions of States; they properly define freedom of expression and sheltered in this right have hacked thousands of institutions around the world, publishing secret information, as a form of struggle against the acts of States; this organization currently works in the form of a network and cannot be regulated because they do not maintain a fixed zone but on the contrary act from all parts of the world (Araujo Torres, 2012).

In short, this is only a general recognition of how this reality has generated an interaction between new and old agents that influence the international system, wherein new cyberspace they have had to adapt to the conditions and in them innovative strategic actions are developed to safeguard the entire global population. It is understood that new acts have been set, which, when implemented, violate treaties or customary rights recognized from the international system. The security and defense of nations, among other acts that respond to the evolution of the conflict and its actions within a new system that continues to reinvent itself and from which it is necessary to continue learning to generate a better containment and defense of possible cyber wars that may be executed, requires an ethical perspective that ensures not to exceed the limits of what is due and legal.

5 Strengthening Security Through the Ethical Use of Data

It is now more often to find State procedures developed by machines and not by humans. Artificial intelligence requires a broad understanding of cybersecurity-related issues given the enormous amounts of sensitive data that the governments use, which could be set as a threat when wrongfully used.

The data provided to decision-makers in the States are tools to become more agile governments, more collaborative, and more connected with citizens to finally improve the quality of life of people (Navas-Camargo, 2020; Cubides-Cárdenas, Navas-Camargo, & González Montes, 2021). The strategic use of data and artificial intelligence in the public sector implies the recognition of great opportunities for the industry in the use of data and technology that Artificial Intelligence implies. Three significant uses have been recognized to support the entire lifecycle of public policy. Combined data and Artificial Intelligence help better understand the situations and problems that cities have, for example, in areas such as mobility, security, public health, and education, among others. On the other hand, Artificial Intelligence helps to provide better services to citizens, assists in designing more personalized services, more timely, and that reaches the right people. It also helps the public administration automate repetitive processes and ensure that civil servants or public servants dedicate themselves to much more analytical and valuable work. Therefore, the aim is to take advantage of this Man–Machine collaboration in the public sector. There are as many opportunities as risks in the use of technologies, especially in the public sector, which is why promoting the ethical and responsible use of information technologies and telecommunications in the public sector has become one of CAF's goals.

CAF (which is the former Andean Development Corporation) is now called the Latin American Development Bank by its Spanish acronym. It is a development bank supporting Latin American countries for 50 years. Until recently, its main emphasis had been on infrastructure-related issues, but since 2018 it created the Directorate of Digital Innovation of the State. It aims to support Latin American governments in the processes of digital transformation.

6 Implementation

The first principle is to identify an organization's ecosystem of tasks and micro-tasks to diagnose whether they can be automated, which others are semi-automated, and which jobs strictly require a human in the whole chain. For a better understanding, an example is set when a public procurement is being pursued. It is essential to understand the difference in prices and the differences between the technical characteristics of a technical request document. A machine could determine the differences in characteristics between a product or service to be contracted and those that exist in the market, analyze the information, and create a report. If the machine can do this, it was previously identified as one of the automatable tasks. Within the life cycle of public procurement, this could apply to justice, public administration, citizen services, security, etc. The second principle is to understand whether it is Big Data or Small Data environments. This is a critical issue for Latin America because people who do Artificial Intelligence are often parameterized to apply all the time machine techniques to understand Big Data, but not so efficient for Small Data. The problems of Artificial Intelligence end up being a problem of excess of data information. Due to the lack of clarity regarding the use that one or another data may have, they generate information bases that are robust in their dimension and vulnerable in their security. Therefore, much attention must be paid to data reengineering and data governance.

Each country must analyze the best way to take advantage of data, govern it, and safeguard it to become a tool that allows the best use, distribution, and access to services and social programs for the citizens.

Artificial Intelligence systems will become an excellent guide for the public policymaker and the implementer of public policies where total transparency is made possible. It will no longer be a person who defines how, when, how much and where to provide a distribution of resources, but on the contrary, a machine that has determined objectively the way to do it.

Colombia has marked a profound leader in trying to deepen the digital transformation from the use of Artificial Intelligence. Other countries such as Uruguay, Peru, and Chile are committed and have launched Artificial Intelligence policies and strategies and guidelines for its implementation, understanding the positive impact that the ethical and responsible use of data generates for governments. However, a phase of the generation of new knowledge is being advanced in the initial stage. It is necessary to find a solution to the problem of information asymmetries and achieve more understanding within public policymakers among public officials in countries where they are not necessarily the epicenters of knowledge origin. A significant challenge will come, and it is the implementation of all these initiatives, deployment of technology, and a massive digital transformation accompanied by a mental change that requires ethical and responsible thinking that allows taking full advantage of the tools, once it is well understood what it is for, what it is helpful for, what it is deficient in, and what it is not inadequate to use.

7 The Case of Colombia

Internet access coverage by the population is still developing. Only 62% of the population has internet access, 49% has Network Availability, and 64% of the country has achieved a satisfactory transition to digitalization. According to the Global Cybersecurity Index-2018, Colombia is ranked 7 out of 333 in the Americas and 73 out of 145 globally.

Recently, at the end of 2019, the National Government of Colombia approved the National Policy for Digital Transformation and AI as part of the National Development Plan (PND) 2018–2022. This included the Pact for the Digital Transformation of Colombia, which set out to consolidate a complete digitalization of the country by 2030. On the one hand, this implies massifying internet connectivity in the personal and organizational environment. Also, to provide the needed resources to assure the development of skills and creative and innovative digital work environments. Finally, the creation of an institution responsible for ensuring the implementation of the Pact and the regulation of new technologies for use in both the private and public sectors.

According to the 2020 report "Artificial Intelligence at the Service of Social Good in Latin America and the Caribbean" by the Inter-American Development Bank (IDB), Colombia is the most competitive country in terms of the efforts made from the entrepreneurial ecosystem and civil society in favor of the implementation of AI strategies, ranking 28th out of 140 countries according to the Global Competitiveness Index of the World Economic Forum in 2018. That same year, the city of Medellin was chosen as the Spanish-speaking capital, home of the Center for the Fourth Industrial Revolution. Colombia is a pioneer in the region in terms of its progress in the private sector, and it's still learning how to make it in the public sector.

In the public sector, its most significant challenge is to face corruption, which leads, among other things, to risks and gaps in terms of cybersecurity. The National Development Plan (NDP) includes a manual for digitization in the local government sphere and a guide to strengthen the achievement of the UN Sustainable Development Goals. The strategy defined by the National Government gave way to the implementation of the Open Data Portal of Colombia that allows access to 10,231 datasets from different government entities, which seeks more transparent access to information. For the sake of the whole strategy, the training of human resources is a fundamental pillar for its success. For this reason, the government granted 25,000 civil servants full scholarships for training in AI and digital transformation topics to be used on the Coursera and Platzi platforms.

Hence the proposal given by CAF is of so much interest. Applying guidelines for better use of AI in the public sector will allow a better use of data and a decrease in the rates of corruption and mismanagement of confidential information and scarce resources. Its main conclusion is that the ethical and responsible use of AI should be based on three pillars, as follows:

1. Define and implement a long-term strategy.
2. A regulatory framework to assist data and algorithmic governance.
3. The human factor. People involved in the procedures are represented not only by those who provide the solution in the public entity but also by those who require it, both in the institution and in the citizenship.

The proposal is just being established, and it will take time to implement it and know how it has been used. Before moving on to a more detailed review of the guidelines, a survey was conducted among 100 people of economically productive age. Most of them lawyers because of their relationship with the authors, one of whom is a lawyer. Answers were provided in several American countries to determine their knowledge of responsible data use and understanding of the regulatory framework set at an organization level and State level. The survey was structured in nine closed-response sections, as follows (Table 1):

The survey was conducted in Spanish. The first three questions were provided to identify the age range of the persons, whether their work was related to the use of sensitive data, and whether it was conducted in the private or the public sector.

The dominant age range was between 39 and 50, and 85% of them answered positively to the 2nd question about the use of sensitive data in their workplace, stating that they do have access to information that is considered sensible (Fig. 1).

As for the knowledge of the long-term strategy defined by their organization to assure a secure and ethical use of the information, 74% of the 100 participants in the survey confirmed they knew about it. From those in the dominant age range, 78% verified their awareness of the policy (Fig. 2).

Along with the given answers and an analysis of them, the CAF proposal is explained to an extent in the following paragraphs.

8 First Guideline—Implementation

Recommendations given by CAF to design and put into work the artificial intelligence strategies need to consider the following parameters:

1. Bring together different actors from the society with leadership in the politics and capacity of coordinating the actions to adopt and implement a public policy

Table 1 Survey questions and answers

Q#	Statement	Option answers	Results
1	Age range, in years	18–28 29–38 39–50 51 or more	9 24 57 10
2	Does your work activity imply that you have access to the use of sensitive personal data, such as identity number and place of residence, among others?	Yes No	80% 20%
3	Is your main work activity in the public or private sector?	Public Private	62% 38%
4	In which country do you work?	Colombia Bolivia Venezuela Mexico Paraguay Panama Dominican Republic United States of America	80% 6% 1% 4% 1% 1% 3% 4%
5	Are you aware of the definition of the long-term strategy that the entity for which you work has defined to ensure ethical and secure treatment of information?	Yes No	74% 26%
6	Do you understand the regulatory framework under which the protection of individuals' information is sought through responsible data governance in your country?	Yes No	33% 67%
7	If you choose Yes in the previous question, what is the name of this regulation?	33 answers were provided	
8	Are you aware of activities to strengthen human capacity, decision-making, and the ethical use of information in your organization?	Yes No	64% 36%
9	Are you aware of activities to strengthen human capacity, decision making, and the ethical use of information in your country?	Yes No	50% 50%

Source Self-made from collected data

for the Artificial Intelligence based on commitments and efforts, coordinated across common goals.

2. Analyze the developments and commitments related to AI and the lessons learned from others. Such as other countries, other organizations, international players, or experts in the field to complement their strategy.
3. Design the strategy and the route with enough flexibility to allow the adoption of new technologies and constant evolution. Also, to implement this strategy,

Fig. 1 57% of surveyed people are in the ages 39–50. 85% of them have access to sensitive data in their workplace. *Source* Automatically made in Microsoft Forms, from the collected data

Fig. 2 In the same dominant group, 78% of the people know about their governing policies. *Source* Automatically made in Microsoft Forms, from the collected data

the realities and necessities of the governments and the sub governments in local communities must be considered.

4. Plan within the strategies governance structures that allow public entities to guide, coordinate, supervise and control what happens during the life cycle of the AI systems.
5. Promote among the public sector a culture that is favorable to considering the pursuit of new ways of acting and analyzing and who are willing to explore and experiment with data. Persons capable enough to adapt themselves in the use and implementation of different methods and to visualize new perspectives off A I without being afraid of taking. Controlled risks. Risk is not a controlled environment.
6. Assure the sustainability of this strategy by making it independent from the political debate and government changes. Being able to continue with the strategy is a crucial element because it determines how efficient and effective the public function can be and its relevance for the country's development.
7. Disseminate the public sector's uses, benefits, and risks associated with AI through an appropriate communication strategy. Generating public confidence is one of the main objectives.
8. Stablished the guidelines to evaluate the technological options when new projects in the public sector are pursued.

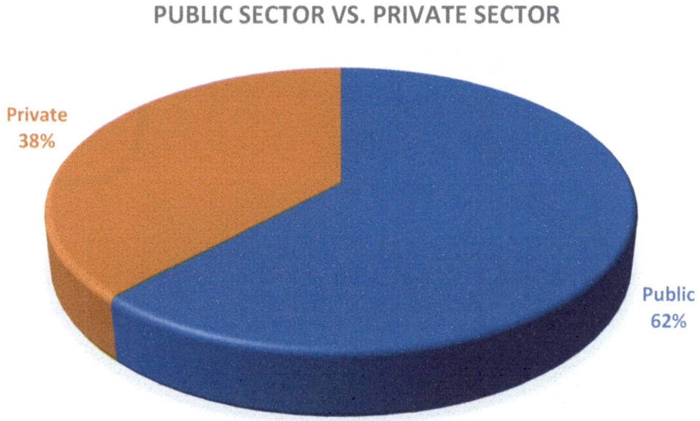

Fig. 3 Percentage of people working in private versus the public sector. *Source* Self-made from collected data

9. Promote the role of the public sector as a driver of entrepreneurship, research, development, and innovation for the improvement of productive processes in the private sector (Fig. 3).

9 Second Guideline—Data Governance

To make the most out of the opportunities offered by AI and to face the challenges associated with it, data and algorithms require good governance and a regulatory frame along with an ecosystem of confidence. In that sense, CAF recommends:

1. Protect the sovereignty of the information. This means that each country must decide where their data is to be kept and reduce the dependence on third parties. For that, guidelines on data storage and processing should be defined and the acceptable level of concentration in a corporation or country.
2. Establish the requirements that data must meet for its quality, completeness, reliability, consistency, and accessibility, among other characteristics. Compliance with the requirements is the only way to ensure that they are helpful for IA tools.
3. Ensure the data value chain. This is achieved by overseeing and evaluating the sequence of processes that transform and add value to the data from its creation to its use and the assessment of its impact.

As seen by the answers provided in the survey, not many people imply to know the regulations related to the data usage, even though most of the participants were lawyers (Fig. 4).

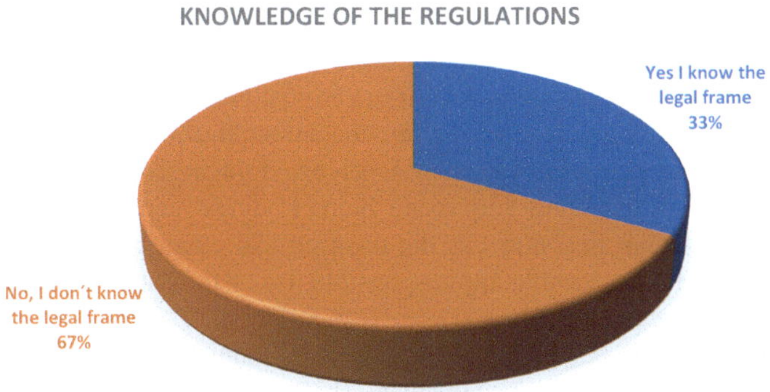

Fig. 4 Knowledge of the regulations

10 Third Guideline—Human Resource

The third element of significant relevance is that the policymakers must always consider that the vital aspect of making better use of AI is the people. This includes the officials who will oversee the development and management of the systems based on this tool and the users, including public officials and citizens. To this end, it is recommended that Latin American governments use as a reference some initiatives implemented by the United States or the United Kingdom for the preparation of human capital, such as:

1. Comprehensive approaches to organizational transformation based on data and AI, with workforce skills development initiatives articulated with the rest of the strategy.
2. Permanent and customized training programs supported by digital platforms, data, and AI.
3. Leveraging data analytics and AI for workforce skills diagnosis, definition, and evaluation of development strategies.
4. Definition of objectives and implementation of strategies in labor welfare, empowerment of workers, and development of an agile and open organizational culture.
5. Definition and implementation of metrics based on key performance indicators to continuously improve the management of workforce readiness for AI adoption.

11 Conclusions

- Globalization has brought into the local agendas many challenges related to their National Security and Defense strategies. Digitalization of information, data, and processes presents as many opportunities as threats.

- One major challenge is that of making responsible use of the collected data. Not only robust machines are needed to prevent a leak of information. But to know why a given Data is being asked, how it can be positively and negatively used, and how it will serve States into shifting into the automatization of certain activities.
- Artificial Intelligence is here to stay. A perceived threat from this kind of technology is having machines surpassing human beings' capabilities. Therefore, focusing on Human Resources as the most critical element when establishing the guidelines for reasonable and responsible use of the data is recognizing the security threats related to AI.
- Colombia is usually being staged because of its many problems and the armed conflict. Nevertheless, and given the many challenges that we still have, the international community has begun to recognize the good deeds of Colombia as being the first Latin American nation to incorporate public policies concerning the usage of Artificial Intelligence in the public service processes.
- The expansion of technologies and interconnection has caused a dependence on them, which is maintained in civil and armed structures such as the public force according to the needs of the State. As a result of this, new designs begin to be implemented to recognize the new space in which disputes can be generated and where it is necessary to establish security, which has become a situation of interest throughout the international community; since the processes have changed to the point at which it is required to make a difference of the possible war scenarios before and after the existence of globalization.

References

Águila, G., Garaño, S., & Scatizza, P. (2016). *Representación estatal y violencia paraestatal en la historia reciente argentina: Nuevos abordajes a 40 años del Golpe de Estado*. Universidad Nacional de La Plata. Facultad de Humanidades y Ciencias de la Educación.

Araujo Torres, M. (2012). *El ciber-espacio internet, sociedad en red y estado (nuevas formas de comunicación, información, y debate)*. Pontifica Universidad Javeriana de Bogotá.

Ardila Castro, C. A. (2018). *La estrategia de ciberseguridad y ciberdefensa en Colombia: una política pública en constante construcción*.

Ardila Castro, C. A., & Cubides-Cárdenas, J. A. (2017). *Política pública de seguridad en Colombia frente a la convergencia y las nuevas amenazas*. Ediciones Escuela Superior de Guerra.

Banco de Desarrollo de América Latina. (2021). *Inteligencia artificial en gobiernos: de la teoría a la acción*. https://www.caf.com/es/actualidad/eventos/2021/06/inteligencia-artificial-en-gobiernos-de-la-teoria-a-la-accion/

Banco Intermaericano de Desarrollo. (2020). *La inteligencia artificail al servicio del bien social en América Latina y el Caribe*. https://publications.iadb.org/publications/spanish/document/La-inteligencia-artificial-al-servicio-del-bien-social-en-America-Latina-y-el-Caribe-Panor%C3%A1mica-regional-e-instant%C3%A1neas-de-doce-paises.pdf

Barlow, J. P. (1996). *Declaración de independencia del ciberespacio*. http://homes.eff.org/%7Ebarlow/Declaration-Final.html

Bitar Giraldo, S. (2006). Cuando los actores no estatales sí importan: el caso de Amnistía Internacional. *Colombia Internacional, 63*, 190–196.

Campillo, B. (2007). *Entre modernidad y globalización*. Pontificia Universidad Bolivariana Medellín.
Castells, M. (2009). *Comunicación y poder*. Alianza editorial.
Cooper, F. (2010). *¿Para qué sirve el concepto de globalización*? http://biblioteca.clacso.edu
Cubides-Cárdenas, J. A., Navas-Camargo, F., & González Montes, L. M. (2021). *El Nuevo Constitucionalismo Latinoamericano*. Direitos Democráticos & Estado Moderno.
Datos Abiertos. (2021). http://herramientas.datos.gov.co
European Commission. (2017). *Cyber diplomacy toolbox*.
Kalyvas, S. (2008). Promises and pitfalls of an emerging research program: The microdynamics of civil war. In: S. Kalyvas, I. Shapiro, & T. Masood (Eds.), *Order, conflict, and violence* (pp. 397–421). Cambridge University Press.
Navas-Camargo, F. (2020). *El sur global y la realidad social de Amércia Latina: hacia la construcción de nuevos paradigmas. Novum Jus, 14*(2), 11–13.
Pana, A. C. (2021). La seguridad cibernética y los derechos humanos-los límites de la restricción de derechos humanos para la protección del espacio cibernético.
Schwab, K. (2017). *The fourth industrial revolution*. Currency.
Sousa Santos, B. (2002). *Hacia una concepción multicultural de los derechos humanos*. Universidad Nacional de Colombia. Instituto de Estudios Políticos y Relaciones Internacionales.
Tzu, S. (2010). *The art of war*. Jaico Publishing House.
Vargas, A. R. (2009) *Guerra civil en colombia: caso de barrancabermeja*. http://www.cerac.org.co/assets/files/guerrayviolencias/9_Guerra_civil_en_Colombia.pdf
Waldmann, P. (1995). *Represión estatal y paraestatal en Latinoamérica*.
Zárate Botía, C. (2008). *Silvícolas, siringueros y agentes estatales: el surgimiento de una sociedad transfronteriza en la Amazonia de Brasil, Perú y Colombia 1880–1932*. Instituto Amazónico de Investigaciones (IMANI).
Zuleta, E. (2015). *Colombia: Violencia, democracia y derechos humanos*. Ediciones Planeta.

Cybersecurity in the Second Decade of the Twenty-First Century

Mariano Bartolomé

Abstract During the second decade of current century, cybersecurity was consolidated as a priority issue of international security. With this framework, our objective is to identify and explain ten cyberattacks and cyberattacks that occurred in the aforementioned period. Through our analysis, we will highlight the heterogeneity of the cybersecurity field, not only in terms of the form of those incidents and attacks, but also the identity of their perpetrators. At the end of the paper, we will highlight the lack of norms and agreements on the international system, to deal with threats and risk from the fifth domain.

Keywords Cybersecurity · Hacking · Computer security · Cybercrime

1 Introduction

Along the second decade of this century, mankind experienced a sustained growth in digitalization processes, which reach and affect every aspect of our lives today. Dependence on Information and Communications Technology (ICT) has increased, and cyberspace has acquired key importance. Here, we understand cyberspace, in a simplified way, as a (mostly) virtual environment of information and interactions among people (Kissinger, 2016). This environment is global and dynamic, supported by ICT infrastructures and systems (Quintana, 2016).

It is correct to consider cyberspace as a "global common", in the way Stang (2013) understands it. That is, a domain that is not under the control or jurisdiction of any State, but its use is a matter of competition between State and non-State actors around the world. But it is important to know that this idea of global common is limited to the Internet infrastructure and its technical aspects, while at the same time there are competitions and power struggles in the actions that take place there (Broeders, 2016).

M. Bartolomé (✉)
Inter American Defense College, Washington, D.C., USA
e-mail: mariano.bartolome@iadc.edu

J. Cayón Peña (ed.), *Security and Defence: Ethical and Legal Challenges in the Face of Current Conflicts*, Advanced Sciences and Technologies for Security Applications,
https://doi.org/10.1007/978-3-030-95939-5_4

The increasing importance of cyberspace in contemporary social activities, from the individual to the global level, has also increased concerns about its security aspects. Thus, from an analytical perspective of cybersecurity influenced by constructivist theoretical elements (Petallides, 2012), the current field is the result of a sustained process of "securitization" (Waever, 1999) of cyberspace started almost three decades ago (Arquilla & Ronfeldt, 1993, 1995; Devost et al., 1995; Schwartau, 1994). Cybersecurity can be broadly understood as focused on the threats that arise and are deployed in cyberspace. A threat presupposes the existence of a hostile actor that can execute them, with the express intention of causing damage. Then, with Rasmussen (2006) we can understand threats as dangers that can be accurately identified and measured, based on the capabilities of the opponent to perform a hostile act (p. 1).

Indeed, both state and non-state actors are active in cyberspace and can carry out offensive actions against a wide range of targets, using different methods and modes. In addition, greater connectivity to Internet also means a greater number of people at risk of attack from cyberspace. As a result of this new situation, cybersecurity was consolidated as a priority issue within the field of international security. In this context, our chapter aims to confirm the important place that cybersecurity occupies in the contemporary international security scenario. To achieve this objective, we will first identify some relevant aspects of the cybersecurity field. Second, we will identify ten major cyber incidents and cyberattacks that occurred during the past decade, describing, and explaining the general characteristics of each of these cases. Finally, in a few brief conclusions, we will confirm that cybersecurity is a heterogeneous and dynamic field that has not yet found effective multilateral institutional responses in the international community.

2 A Few Words About Cybersecurity

The concept of cybersecurity has different meanings and limits, depending on the source employed. In fact, Maurer and Morgus (2014) have compiled more than fifty different definitions of cybersecurity. To overcome this diversity, here we employ the broad and inclusive definition proposed by the United Nations through ITU (2018: 13):

> Cybersecurity is the collection of tools, policies, guidelines, risk management approaches, actions, trainings, best practices, assurance and technologies that can be used to protect the availability, integrity and confidentiality of assets in the connected infrastructures pertaining to government, private organizations and citizens; these assets include connected computing devices, personnel, infrastructure, applications, services, telecommunications systems, and data in the cyber-environment.

The ITU definition indicates that the core of cybersecurity is the protection of the availability, integrity and confidentiality of assets against possible attacks or malicious activities. These aggressions and malicious activities have multiple forms, so it is useful to adopt additional criteria to understand them. A useful approach, used both in academia (The Hague Centre for Strategic Studies, 2015) and among ICT companies (McKay et al., 2015; Microsoft, 2013), is to classify these events according to four elements: actors or perpetrators, targets, actions, and impacts. The actor is the entity that wishes to use cyberspace for malicious activities. The target is the entity that can be subject to those activities. The actions are the tools and techniques to use in a malicious activity. Finally, the impact refers to the damage of the cyber malicious activity.

We have already indicated that the actors that can execute aggressions in cyberspace are extremely heterogeneous and can be state or non-state actors. Specialized literature (Burton, 2015; Lindstrom, 2012) identifies, among the most relevant non-state actors, terrorist or criminal groups, espionage organizations, private companies, hackers, and so-called "insiders". Important and dangerous phenomena that are of priority importance in the field of cybersecurity are associated with some of these actors, such as cyber terrorism, cybercrime y cyber espionage.

In recent years, non-state groups with significant cyber capabilities have become increasingly important as actors, carrying out complex and extensive cyber espionage activities, as well as cybercrime actions. These groups often operate as "proxies" for some nation states and are called Advanced Persistent Threat (APT). Then an APT is a group which "uses continuous, clandestine, and sophisticated hacking techniques to gain access to a system and remain inside for a prolonged period of time, with potentially destructive consequences" (Kaspersky, s/f).

Nation states can also be involved in this type of activity, through civilian or military agencies, using their "cyber power" in different ways. Cyber power can be understood, broadly, as "the ability to use cyberspace to create advantages and influence events in all the operational environments and across the instruments of power" (Hathaway & Klimburg, 2012: 28). It also consists in "the ability to obtain preferred outcomes through use of the electronically interconnected information resources of the cyber domain" and involves both *hard power* and *soft power* formats (Nye, 2010: 4).

Regarding the actions, tools and techniques used in the malicious activities, they are varied. As we have indicated in previous works (Bartolome, 2020), these actions have a set of characteristics that increase their danger. In addition, malicious actions in cyberspace tend to be relatively low cost compared to other forms of damage generation. In this sense, the "weaponry" used by the aggressor may consist of everyday computers and affordable software, while the executor does not require very sophisticated knowledge. Over time, the knowledge required to perform a malicious action in cyberspace, and its sophistication, have constantly decreased, while the effectiveness of the act, and the damage generated, have increased (Gaidosh, 2018; Robinson, 2016).

In cybersecurity the range of tools and techniques is enormous, precisely because of the scope of these concepts, and the evolution of technological development. In this context, the so-called "malware", malicious software specially designed to be downloaded to, or introduced in, a computer, where it can cause serious damage or data breaches, occupies an important place. The most known forms of malwares are *viruses* (spreads with your action), *worms* (spreads automatically), *trojans* (disguised as legitimate software), *spyware* (monitors the computer activity) and *ransomware* (blocks access to the files or computer). Also widely used, with high success, is the Distributed Denial of Service (DDoS): the intentional paralyzing of a computer network by flooding it with data sent simultaneously from many computers. Thousands of botnets, which are connected devices that a malware turns into a bot, are used to execute a DDoS.

In terms of tools and techniques, we must remember the use of the *Deep Web*, which is that huge portion of the Internet whose sites are not indexed in traditional search engines. Especially the so-called *Dark Web*, where privacy is the central element, and consequently the information is encrypted. This portion of the Internet is exploited for lawless practices and businesses, which are paid with cryptocurrencies that are very difficult to trace. The Dark Web poses the challenge of finding a balance between freedom of information and privacy, on the one hand, and malicious cyber activities, on the other (Kumar & Rosenbach, 2019).

An analysis of malicious activities in cyberspace that does not focus on the actors, or on the tools and techniques, but on the type of targets, should pay special attention to the so-called "critical infrastructures". This designation refers to infrastructure and assets that are of vital importance to national security, governance, public health, and the economy, and to public trust. Basically, the idea of critical infrastructures refers to systems, machines, buildings, or facilities related to the provision of essential services to the population (Quintana, 2016: 95). These infrastructures include the information processing and telecommunications systems, the software to operate them, and the personnel who operate those systems and use that software.

To conclude this section of the chapter, another approach for rating cyber-aggressions or malicious activities in cyberspace evaluates their intensity. These offensive acts can fluctuate, across a wide spectrum, between "cyber incidents" and "cyberattacks" of varying severity. Cyber incidents consist of security events that compromise the integrity, confidentiality, and availability of an information asset (The Hague Centre for Strategic Studies, 2015). Cyberattacks, on the other hand, are malicious actions aimed at collecting, disrupting, denying, or destroying information system resources, or information itself (Idem). Due to their increased intensity, cyber-attacks can cause injury or death to people, as well as damage or destruction to objects (Schmitt, 2017). The possibility of executing cyberattacks is conditioned by the possession of appropriate "cyberweapons", which Stevens (2018: 483) describe as "computer code that is used, or designed to be used, with the aim of threatening or causing physical, functional, or mental harm to structures, systems, or living beings".

Because of their intensity and their effects, cyberattacks can constitute a National Security issue, and for this reason National Cyber Security consists in:

> The focused application of specific governmental levers and information assurance principles to public, private and relevant international ICT systems, and their associated content, where these systems directly pertain to national security. (Hathaway & Klimburg, 2012: 28)

National Cyber Security aims to achieve a secure environment in which citizens can be protected from various cyber-attacks, and related non-cyber attacks, both domestic and foreign. It encompasses governmental, national and international dimensions and can be interpreted from different perspectives, each with different priorities (Idem).

3 The Second Decade of the Twenty-First Century

In 2010 there was a paradoxical situation related to cyberspace, which reflected the coexistence of two contradictory perspectives on the situation in that environment. From an Italian scientific publication, the Internet was officially nominated for the Nobel Peace Prize, for being a medium that encourages dialogue, debate and consensus at a global level, through communication. This idea was supported by numerous intellectuals, scientists, and artists, and was accepted by the organizers of the prize. But, at the same time, prestigious journalistic and academic media, with worldwide prestige, confirmed that this environment had been transformed into a new "domain" of war. The fifth domain, in addition to the traditional land, sea, air and space domains. "Warfare has entered the fifth domain: cyberspace", said The Economist (2010), while William Lynn, the United States Deputy Secretary of Defense, stated:

> As a doctrinal matter, the Pentagon has formally recognized cyberspace as a new domain of warfare. Although cyberspace is a man-made domain, it has become just as critical to military operations as land, sea, air, and space. (Lynn, 2010)

The public designation of cyberspace as the fifth domain reflected the importance that this dimension of human activity has attained, not only in military matters, but also in terms of security in a broader sense. The importance of the fifth domain was based on different events that had occurred in previous years, especially two that involved Russia and occurred in Estonia (2007) and Georgia (2008). The Estonian case posed a major dilemma for the North Atlantic Treaty Organization (NATO), to which the small Baltic nation belonged, regarding the type of response it should adopt. As the Organization did not yet have a clear position on cybersecurity, it was unclear whether the cyberattack fell under the collective defense mechanism provided for in Article 5 of the North Atlantic Treaty. That article states that an armed attack against one or more allies shall be considered as an attack directed against all members. To fill this important gap, NATO officially approved its first Cyber Defense Policy in early 2008 (Theiler, 2011).

Moreover, this importance is confirmed by some statements made by influential academics and government officials on this subject. For example, earlier in the decade, Richard Clarke -who supervised the U.S. counterterrorism office during the 9/11 attacks—indicated that "bits and bytes" could destroy physical infrastructure as effectively as a kinetic weapon (Clarke & Knake, 2011). Joe Lieberman, Chairman of the Senate Homeland Security Committee, spoke about the risks of a "Cyber 9/11", while Leon Panetta, Secretary of Defense, warned that the superpower faced the possibility of a "Cyber Pearl Harbor," as destructive as the terrorist attacks on the Twin Towers and the Pentagon (Buchanan, 2020).

Such views were commonplace over the following years. Paradoxically, this pessimistic outlook was not reflected in the signing of a comprehensive treaty on cybersecurity with global scope and high consensus, such as a convention at the United Nations. Even today, this negative situation is related to the different interpretations of this issue among the main nation-states (Gold, 2019; Urgessa, 2020). As anticipated at the beginning of this text, in Western democracies, cybersecurity focuses on the availability, integrity, and confidentiality of information. On the other hand, in China, Russia and other countries with authoritarian or totalitarian regimes, cybersecurity also extends to the contents, which can be controlled and censored, since information is understood as a weapon, which can be employed through cyberspace. Thus, these incompatible positions are the main reasons for the lack of global-level agreements on the issue (Tsaruk & Korniiets, 2020).

Nothing changed by the end of the decade. For example, David Sanger called cybernetics "the perfect weapon" because of its adaptability, low cost, and anonymity, which allows total deniability on the part of the perpetrators. Furthermore, this three-time Pulitzer Prize winner argued that cyberweapons enable a wide range of offensive operations and have displaced terrorism and nuclear attacks as the greatest threat to global security (Sanger, 2018). Indeed, opinions like Sanger's were not isolated events. The World Economic Forum, in its 2019 report, included cyberattacks in its Top Ten Risks, ranking #5 in terms of likelihood, and #7 in terms of impact (World Economic Forum, 2019). The insurance company Allianz, in its 2019 Risk Barometer, ranked cyber incidents as the second most important business risk globally, and in the first place in the following year (Allianz Global Corporate & Specialty, 2020). By 2021, cyber incidents shared the place of the most important business risks globally, along with a pandemic outbreak and business interruption (Allianz Global Corporate & Specialty, 2021).

The review of ten cybernetic events that occurred in the second decade of the century confirms the high importance achieved by these issues in the field of International Security. The following series of selected events presents the different forms that cyber-incidents and cyber-attacks can take nowadays.

4 Study Cases

4.1 Stuxnet

That was the name assigned to a virus hacked into the computer system of Iran's Natanz nuclear power plant in early 2010. The virus, a cyberweapon, entered the plant via an infected USB memory stick and spread through the computers, until it found the software of the machines used to enrich uranium. From that moment on, it irreparably damaged hundreds of these machines, which could be the key element for the development of nuclear weapons. Until today, the predominant interpretation of the case is that the virus was created by the United States and Israel to sabotage

Iran's nuclear program. This version is officially supported by the regime in Tehran and even by American journalists and is also accepted in the academic community. Obviously, U.S. and Israeli authorities have never confirmed or denied this claim. Symantec computer security company issued a comprehensive report on Stuxnet and indicated that its complexity was so high that few players in the world had the capacity to develop it. At the end of the report, Symantec concluded that "Stuxnet is the type of threat we hope to never see again" (Falliere et al., 2011: 55).

Stuxnet case has been described as a preemptive attack, employing offensive cyber capabilities although in a defensive sense (Smeets & Lin, 2018). Instead, other approaches labeled it as a "degradation" cyberattack, a type of attack that generates high costs and damages, aimed at destroying or degrading the opponent's critical capabilities. From that point of view, Stuxnet was highly effective, although it also showed the limited coercive power of the cybernetic tool, as Iran did not abandon its nuclear weapons plan (Maness & Valeriano, 2018).

4.2 Wikileaks

Also, in 2010 the affair known as Wikileaks happened. Wikileaks is a web page created by the Australian hacker Julian Assange to disseminate official or corporate U.S. documentation, especially corruption cases, that was not releasable to the public, always preserving the source. At that time, this web page leaked more than 700,000 classified Pentagon and State Department documents on the wars in Afghanistan and Iraq. The objective was to show cases of violation of individual freedoms, of International Humanitarian Law and even of Human Rights, by state agencies in the context of the global war on terrorism.

The exfiltrated information was disseminated quickly and massively by the world's leading newspapers. This was described as one of the largest leaks of classified information in U.S. history. The Federal Government's investigations accused a former soldier and Army intelligence analyst as the "insider". That person possessed the security clearance necessary to access the classified documents and communications which were leaked under Assange's coordination.

This case generated numerous debates about the ethics of state management of information and citizens' rights to access it, which are beyond the scope of this paper. Beyond these counterpoints, the Wikileaks affair demonstrated, as indicated in the most important book on the issue (Sifry, 2011), the capacity of a non-state organization in the digital era to hamper the will of the most powerful governments to safeguard information. Somehow, the White House tacitly admitted these potentialities by considering Wikileaks as a sort of "hostile non-state intelligence service" (Bartolomé, 2020).

4.3 Anonymous

In 2012, in the traditional poll organized by TIME magazine, Anonymous was voted the first of the 100 most influential personalities in the world. Anonymous refers to a global network of hacktivists, without clear leadership, who coordinate their actions to carry out mass protests or claims on the web. This network has carried out cyberattacks against government sites, certain corporations and even terrorist groups (including ISIS) on numerous occasions. The focus of their protests is usually set on the inequalities of the globalization process, consumer rights, freedom of expression, human rights, and the violation of citizens' privacy by governments.

Their attacks consist of hacking servers, stealing sensitive data, and disseminating it, or blocking it via DDoS. The most publicized action took place in 2008, when Anonymous articulated two waves of global protests against the so-called Church of Scientology. Through this event, the network demonstrated the planetary reach of its actions and its impact on public opinion. The influence of Anonymous confirmed the effectiveness of appropriation and disclosure of sensitive information as a form of political activism in current times (Torres Soriano, 2013).

4.4 Sony Pictures

In late 2014 the film company Sony Pictures was the target of a cyberattack, allegedly perpetrated by a group of hackers called Guardians of Peace (GOP). GOP justified the aggression as retaliation for the film The Interview, a satire of North Korean leader Kim Jong-un, being made by the company. In addition, the aggressors intended to stop the dissemination of the film. To achieve this goal, they claimed that the reprisal would be extended to cinema operators who screened the film, and therefore many of them decided not to do so. The attack consisted of the theft of information from the company and its employees, including thousands of e-mails from its top executives, which was leaked to the public in the following weeks. Entire films already produced but not yet presented to the public were also stolen. In addition, in the months following the cyberattack, the company reportedly lost close to US$ 200 million due to the loss of value of its shares in the stock exchanges, loss of profits due to delays on the opening day and financial compensation to employees whose personal data were leaked.

Federal Bureau of Investigation (FBI) inquiries concluded that the malware used to steal information from Sony's computer network was the same to the one used in March 2013 by North Korea-linked hackers in a similar attack on South Korean banks. Therefore, FBI identified the Pyongyang regime, which months earlier had called the film an act of war and a manifestation of terrorism, as responsible. This case was described by the White House as a matter of National Security, even though it only involved a private company, whose capital is not wholly owned by the United States. This interpretation was justified by the White House because of the intention of Kim regimes to censor the right of American citizens to express themselves with freedom (Murphy & Scanell, 2014). Moreover, it was the first relevant episode of a trend that later became clearer: the growing capacity of hackers overcomes the efforts of the private sector, even of the largest companies, to preserve their information infrastructure. For this reason, the collaboration of the State in the design and implementation of effective solutions is necessary (Simon, 2017).

4.5 *Black Energy*

The Black Energy case refers to a malware (a cyberweapon) of the same name used to degrade Ukraine's power generation grid, causing massive power blackouts hours before Christmas 2015 in the west of the country. The decision to launch the cyberattack was attributed to Russia, which was involved in a major diplomatic crisis with the neighboring nation. The malware entered the power grid through a phishing[1] attack, carried out a few months earlier, damaged the control centers of several stations and left a quarter of a million inhabitants without service for several hours. After different analysis, the responsibility was attributed to Sandworm Team, an APT type group linked to Russia that had already used a previous version of this software in the 2007 attacks on Georgia (Park & Walstrom, 2017). Although Black Energy was a lesser known event than others of a similar nature, one author considers it a kind of "turning point" in cyberattacks, because it was a deliberate attack on a critical infrastructure for civilian use (Buchanan, 2020).

4.6 *Russiagate*

After the Black Energy case, Russia was once again in the eye of the storm with the so-called Russiagate. This is the name given to the actions supposedly taken by the Putin government in connection with the 2016 U.S. presidential election. These actions were of two types: espionage and disinformation operations. The espionage consisted of hacking into the Democratic National Committee and accessing the emails of its members, via phishing, by Fancy Bear (APT-28). This group is linked to Russian intelligence and has been accused of carrying out numerous previous actions against high-value public targets in different countries (Fire Eye, 2017).

The disinformation operations were designed in, and executed from, the Internet Research Agency (IRA). This institution had headquarters in St. Petersburg and, despite its allegedly independent and non-partisan position, operated according to the guidelines of the country's Executive branch (Mac Farquhar, 2018; Sampedro Oliver, 2021). Russian interference in the electoral process consisted in the design

[1] Phishing: messages used by scammers to trick the victim into clicking a link or an attachment that will provide them access to information or download malware onto the computer.

and dissemination of thousands of political advertisements through hundreds of fake accounts mainly on Facebook, but also on Twitter and Instagram, as well as the use of the Google platform (CCN-CERT, 2019; Redondo, 2017). Two reports prepared by the U.S. Congress within the investigations carried out by its Intelligence Committee indicated that some twenty Facebook pages managed by IRA were visited by 126 million people and their contents obtained 39 million "clicks" of approval ("likes"). Another 20 million people visited the Instagram pages managed by this Russian Agency (Timberg & Romm, 2018).

4.7 *Equifax 2017*

Equifax is an important and well-known credit analysis company located in the United States, which was the victim of a breach of its files by a cyber espionage group. The intrusion began in March 2017 and lasted almost three months until it was discovered. Initially, three servers were breached, which allowed intruders to access another half a hundred. In this way, the attackers gained access to the personal data (IDs, SSNs, credit card, medical information, addresses, etc.) of almost 148 million people; that is, more than half of the population of the United States. The scale of the attack was so large, that it prompted Congress to push for a government investigation in addition to a judicial one (US GAO, 2018). Three years later, federal prosecutors filed charges against several alleged members of the Chinese People's Liberation Army (PLA) involved in the attack, described as a deliberate and "sweeping" intrusion into the private information of the people of the United States (Department of Justice, 2020).

4.8 *Wannacry-NotPetya 2017*

In 2017, two major cyber incidents occurred on a global scale, spreading through worms: WannaCry and NotPetya ransomwares. WannaCry exploited flaws in Microsoft Windows operating systems, is attributed to the North Korea-linked "Lazarus" APT-type group and is estimated to have infected millions of computers in some 150 countries. In fact, Europol described it as a cyber-attack of global scale and dimension never seen before (BBC, 2017). The attackers demanded a monetary "ransom" for the release of each controlled computer, which varied depending on the status of the target, but increased over time. Of course, there are no exact figures of the amount of money made by the attackers, but some estimates speak of millions of dollars. The targets were public and private, from different economic sectors, and many of them were critical infrastructures. Some of the targets were the railroad companies of Russia and Germany; the National Health Service of the United Kingdom; Telefónica of Spain; the automobile companies Honda (Japan) and Renault (France); Latam airlines; the Judiciary of Sao Paulo, Brazil; the Ministry of Security of China; and the private courier company Fedex of the United States (Deloitte, 2017).

Shortly after WannaCry, NotPetya was detected in Ukraine, where it infected government offices, the Kiev airport, and different banks, among other targets. In a second stage, NotPetya affected multiple targets in more than 75 countries. Among them, the container companies Maersk (Denmark) and TNT (Netherlands), affecting global maritime trade flows; the U.S. pharmaceutical company Merck; and the Russian oil company Rosneft. It has been estimated that the damages and losses generated by the attackers, who demanded the ransom in bitcoins, could have reached almost US$ 10 billion (Banerjea, 2018). Until today, responsibility for this attack has never been established with any certainty, however, the U.S. government blamed Russia, through its military intelligence agency GRU (Ibid.; Nakashima, 2018).

4.9 Cambridge Analytica

In 2018, after an in-depth journalistic investigation (The New York Times, working with The Observer of London and The Guardian), the Cambridge Analytica scandal broke out, which had occurred four years earlier. This British political marketing and public opinion company, without authorization, used personal data on preferences and interests of 270,000 direct Facebook users, and another 50 million indirectly. It did so through the "This is your digital life" application, presented to Internet users as a social research tool. The data collected in this way by the company was used to create voter profiles, which were sold for use in political campaigns, including Donald Trump and Brexit (Confessore, 2018). The Cambridge Analytica affair, coupled with Russiagate, generated enormous concern about the unauthorized use of personal data that exist on social networks for targeted advertising. Specialized reports indicated that the operating model of Facebook, other social networks and the Google platform deeply erodes the right of privacy, facilitates the manipulation and influence on users and, in the end, constitutes a systematic threat to their human rights (Amnesty International, 2019).

4.10 Solar Winds

Closing our list of selected events, and the review of the decade, in December 2020 a cyber-attack of vast dimensions was discovered, carried out by a group of hackers who were breaching targets around the world and performing massive information theft activities. The investigations found that tens of thousands of organizations were successfully attacked, including most of the Fortune 500 companies and several U.S. federal agencies, including: the Office of the President of the United States, the Departments of State, Treasury, Commerce and Homeland Security, the National Institute of Health, the Pentagon and the Department of Energy, including the National Nuclear Security Administration. The attack reportedly began in March and essentially involved the insertion of malicious code (called Sunburst) into the Orion software developed by Solar Winds, a Texas-based company that sells cybersecurity software. When users of this software installed it, they also downloaded the malicious code that opened a "backdoor", granting access to the hackers.

"This can turn into one of the most impactful espionage campaigns on record," said cybersecurity expert Dmitri Alperovitch (Tucker et al., 2020). The severity of this event prompted the formation of a Unified Cyber Coordination Group (UCG) to coordinate an integrated, government-wide response. The Group was composed of the Federal Bureau of Investigation (FBI), the Cybersecurity and Infrastructure Security Agency (CISA), and the Office of the Director of National Intelligence. The FBI investigated and collected intelligence to attribute, pursue and interdict the actors responsible for the threats. CISA instructed federal civilian agencies to immediately disconnect or shut down the affected Solar Winds Orion products from their network. The Agency also provided technical assistance to those affected by the attack, helping them to recover quickly (ODNI, 2020). Although the U.S. government never made an official attribution, officials admitted that the perpetrator was likely of Russian origin. Weeks later, senators confirmed, based on preliminary reports, that the scope, scale and implications of this incident were greater than any previously faced by the United States (Diaz Cardel, 2021).

5 Conclusions

Our analysis allows us to conclude that, during the second decade of the century, security issues in the cyber domain have steadily increased in importance. Despite the nomination of the Internet for the Nobel Peace Prize due to its positive contribution to communication and dialogue between people globally, cyberspace has been the scene of numerous risks and threats of different type. Thus, it can be established that in the period under consideration, the field of cybersecurity has been characterized by its intensity and a high degree of heterogeneity.

This heterogeneity can be seen not only in the format of the cyber incidents and cyberattacks described and explained above, but also in the nature of their protagonists. In most of the cases (Stuxnet, Sony Pictures, Black Energy, Russiagate, Equifax, WannaCry/NotPetya and Solar Winds), the investigations allow us to consider that different nation-states had a high degree of responsibility in their direction, and probably also in their execution. In none of these seven cases could state responsibility be proven in a conclusive and unappealable manner, demonstrating the complexity of "attribution" in cyberspace.

The existence of APT-type groups that at some times act autonomously, and at other times become state proxies, contributes to this complexity. Although APT groups generally fall under the label of cyber espionage activities, the pursuit of financial gain through extortive acts, as in WannaCry and NotPetya, also coincides with a cybercrime profile. In addition to nation-states and APT-type groups, the list of actors involved in cybersecurity events during the decade under study should also include hacktivist movements (Anonymous) and insiders (Wikileaks), with an extraordinary capacity for generating damage.

The targets have been state-owned (Stuxnet, Black Energy, Wikileaks), private (Sony Entertainment, Equifax, Anonymous), mixed (Solar Winds, WannaCry, NotPetya, Anonymous) and even civil society as an actor, as in the case of Cambridge Analytica and Russiagate. Critical infrastructures have confirmed their high value in terms of cybersecurity. The cases of Stuxnet and Black Energy have been hostile actions aimed directly at such targets, specifically in the energy sector, although only the latter is of civilian use.

WannaCry and NotPetya also affected such targets, although not in a premeditated manner, but by global-scale replication of the malware. All but three of the cases studied (Anonymous, WannaCry and NotPetya) affected the national security of the countries where they occurred, in one way or another. This was true even in the case of Sony Entertainment, a privately held multinational company. At the same time, none of them was approached in a military key, confirming that a modern conception of National Security is not limited to the field of *hard power*.

The tools and techniques used by the perpetrators of the cyber incidents and cyber-attacks analyzed in this chapter have been diverse. Among them, malware, viruses (e.g., Stuxnet and Black Energy) and also worms (e.g., WannaCry and NotPetya) . In the latter two cases, they have also been ransomwares. To add further elements to this diversity, Anonymous has frequently used DDoS, while the phishing technique has undoubtedly been used on at least three occasions (Black Energy, Russiagate and Sony Pictures).

All this complexity confirms that cybersecurity occupies a privileged position on the contemporary international security agenda. It is foreseeable that the hierarchy of cybersecurity will tend to increase in the short and medium term, due to the dynamics observed in cyberspace because of technological progress. However, this importance is not being reflected in the very limited consensus reached on these issues at the level of multilateral institutions, beyond merely declamatory attitudes with no real backing. There is no doubt that the signing of a Convention on Cybersecurity, with strong support from the main state actors, would contribute decisively to a more effective treatment of these issues. However, the profound differences in these matters between Western democracies and countries such as China and Russia do not help to be optimistic. An increase in governance mechanisms around different aspects of cybersecurity on a global scale will also have a strong positive impact on the field. In an ideal scenario, these mechanisms should actively involve a variety of public, private, and civil society stakeholders. In short, the challenge in the short and medium term is to achieve greater and better institutionalization.

References

Allianz Global Corporate & Specialty. (2020). *Allianz risk barometer. Identifying the major business risks for 2020*. Allianz Global Corporate & Specialty.

Allianz Global Corporate & Specialty. (2021). *Allianz risk barometer. Identifying the major business risks for 2021*. Allianz Global Corporate & Specialty.

Amnesty International. (2019). *Surveillance giants: How the business model of Google and Facebook threatens human rights*. POL 30/1404/2019. Amnesty International.

Arquilla, J., & Ronfeldt, D. (1993). Cyberwar is coming! *Comparative Strategy, 12*(2), 141–165.

Arquilla, J., & Ronfeldt, D. (1995). Cyberwar and netwar: New modes, old concepts, of conflict. *RAND Research Review, XIX*(2).

Banerjea, A. (2018, August 27). NotPetya: How a Russian malware created the world's worst cyberattack ever. *Business Standard*.

Bartolome, M. (2020). Las Ciberamenazas y su impacto en el campo de la Seguridad Internacional. *Revista de la Escuela Superior de Guerra, 602*, 151–163.

BBC. (2017, 12 de mayo). El ciberataque de escala mundial y "dimensión nunca antes vista" que afectó a instituciones y empresas de unos 150 países. *BBC Mundo*.

Broeders, D. (2016). The public core of internet: Towards an international agenda for internet governance. *CyFy Journal, 3*, 24–30.

Buchanan, B. (2020, February 2). Five myths about cyberwar. *The Washington Post*.

Burton, J. (2015). NATO's cyber defence: Strategic challenges and institutional adaption. *Defense Studies*, 1–22.

CCN-CERT. (2019, febrero 13). *Desinformación en el ciberespacio*. CCN-CERT BP.

Clarke, R., & Knake, R. (2011). *Cyber war: The next threat to national security and what to do about it*. Harper Collins.

Confessore, N. (2018, April 4). Cambridge Analytica and Facebook: The scandal and the fallout so far. *The New York Times*.

Deloitte. (2017, junio). *¿Qué impacto ha tenido el ciberincidente de WannaCry en nuestra economía*? Deloitte Cyber Risk.

Department of Justice. (2020, February 10). *Chinese military personnel charged with computer fraud, economic espionage and wire fraud for hacking into credit reporting agency Equifax*. Office of Public Affairs.

Devost, M., Houghton, B., & Pollard, N. (1995). *Information terrorism: Can you trust your toaster*? (Sun Tzu Art of War writing award, National Defense University). The Terrorism Research Center.

Diaz Cardel, J. (2021, 26 de febrero). ¿Está perdiendo EE.UU. la guerra tecnológica y de ciberseguridad frente a Rusia y China? *Escudo Digital*.

Falliere, N., Murchu, L., & Chien, E. (2011, February). *W32.Stuxnet Dossier. Version 1.4*. Symantec Security Response.

Fire Eye. (2017, January). *APT28: At the center of the storm*. Special Report.

Gaidosh, T. (2018, June, 22–25). The industrialization of cybercrime. Lone-wolf hackers yield to mature businesses. *Finance & Development, 55*.

Gold, J. (2019, May 16). Two incompatible approaches to governing cyberspace hinder global consensus. *Leiden Security and Global Affairs*.

Hathaway, M., & Klimburg, A. (2012). Preliminary considerations on national cyber security. In A. Klimburg (Ed.), *National cyber security framework manual* (pp. 1–43). NATO CCD COE.

International Telecommunications Union. (2018). *Guide to developing a national cybersecurity strategy. Strategic engagement in cybersecurity*. ITU.

Kaspersky. (s/f). What is an advanced persistent threat (APT)?

Kissinger, H. (2016). *Orden Mundial*. Debate.

Kumar, A., & Rosenbach, E. (2019, September 22–25). The truth about the dark web. *Finance and Development, 56*(3).

Lindstrom, G. (2012). *Meeting the cyber security challenge*. Geneva Centre for Security Policy. Geneva Papers 7-2012.

Lynn, W. (2010). Defending a new domain. The Pentagon's cyberstrategy. *Foreign Affairs, 89*(5), 97–108.
Mac Farquhar, N. (2018, February 18). Inside the Russian troll factory: Zombies and a breakneck pace. *The New York Times.*
Maness, R., & Valeriano, B. (2018). International cyber conflict and national security. In D. Reveron, N. Gvosdev, & J. Cloud (Eds.), *The Oxford handbook of US national security* (pp. 403–419). Oxford Handbook Online.
Maurer, T., & Morgus, R. (2014). *Compilation of existing cybersecurity and information security related definitions.* Federal Department of Foreign Affairs.
McKay, A., et al. (2015). *International cybersecurity norms reducing conflict in an internet-dependent world.* Microsoft Corporation.
Microsoft. (2013). *Five principles for shaping cybersecurity norms.* Microsoft Corporation.
Murphy, M., & Scanell, K. (2014, December 18). Sony hack a 'National Security Matter'. *Financial Times.*
Nakashima, E. (2018, January 12). Russian military was behind 'NotPetya' cyberattack in Ukraine, CIA concludes. *The Washington Post.*
Nye, J. (2010, May). *Cyber power.* Belfer Center for Science and International Affairs.
ODNI. (2020). Joint statement by the Federal Bureau of Investigation (FBI), the Cybersecurity and Infrastructure Security Agency (CISA), and the Office of the Director of National Intelligence (ODNI). ODNI News Release No. 44-20, December 16.
Park, D. & Walstrom, M. (2017, October 11). Cyberattack on critical infrastructure: Russia and the Ukrainian power grid attacks. University of Washington, The Henry M. Jackson School of International Studies. *JSIS News.*
Petallides, C. (2012). Cyber terrorism and IR theory: Realism, liberalism, and constructivism in the new security threat. *Inquiries Journal/Student Pulse, 4*(03).
Quintana, Y. (2016). *Ciberguerra.* Ediciones de la Catarata.
Rasmussen, M. (2006). *The risk society at war: Terror, technology and strategy in the twenty-first century.* Cambridge University Press.
Redondo, M. (2017, 2 de noviembre). Publican los anuncios comprados por Rusia en Facebook. *Hipertextual.*
Robinson, N. (2016, June 8). NATO: Changing gear on cyber defense. *NATO Review.*
Sampedro Oliver, R. (2021, de marzo 9). Redes sociales: desinformación, adicción y seguridad. Instituto Español de Estudios Estratégicos (IEEE). *Documento de Opinión.* 30/2021.
Sanger, D. (2018). *The perfect weapon. War, sabotage and fear in the cyber age.* Crown Publishing.
Schmitt, M. (Ed.). (2017). *Tallinn manual 2.0 on the international law applicable to cyber warfare.* Cambridge University Press.
Schwartau, W. (1994). *Information warfare: Chaos on the electronic superhighway.* Thunder Mouth Press.
Sifry, M. (2011). *Wikileaks and the age of transparency.* Yale University Press.
Simon, D. (2017, October). *Raising the consequences of hacking American companies.* Center for Strategic and International Studies (CSIS).
Smeets, M., & Lin, H. (2018). Offensive cyber capabilities: To what ends? In T. Minarik, R. Jakschis, & L. Lindstrom (Eds.), *10th International Conference on Cyber Conflicts: Maximizing Effects* (pp. 55–72). NATO CCD COE.
Stang, G. (2013, April). Global commons. Between cooperation and competition. European Union Institute for Security Studies. *Issue Brief* No. 17.
Stevens, T. (2018). Cyberweapons: Power and the governance of the invisible. *International Politics, 55*, 482–502.
The Economist. (2010, July 3). War in the fifth domain. Are the mouse and keyboard the new weapons of conflict? *The Economist.*
The Hague Centre for Strategic Studies. (2015). *Assessing cyber security. A meta-analysis of threats, trends, and responses to cyber attacks.* The Hague Centre for Strategic Studies.

Theiler, O. (2011, de septiembre 11). Nuevas amenazas: el ciberespacio. *Revista de la OTAN. Edición digital.*

Timberg, C., & Romm, T. (2018, December 17). New report on Russian disinformation, prepared for the senate, shows the operation's scale and sweep. *The Washington Post.*

Torres Soriano, M. (2013, de diciembre 10). Siete lecciones no aprendidas sobre Anonymous. Instituto Español de Estudios Estratégicos. *Documento Opinión.* 122/2013.

Tsaruk, O., & Korniiets, M. (2020). Hybrid nature of modern threats to cybersecurity and information security. *Smart Cities and Development Journal, 4*(11), 57–78.

Tucker, E., Bajak, F., & O'Brien, M. (2020, December 13). US agencies hacked in monthslong global cyberspying campaign. *AP News.*

United States Government Accountability Office. (2018, August). *Data protection. Actions taken by Equifax and Federal Agencies in response to the 2017 breach.* GAO-18-559.

Urgessa, W. (2020). Multilateral cybersecurity governance: Divergent conceptualizations and its origin. *Computer Law & Security Review, 36*, 1–8.

Waever, O. (1999). Securitization and desecuritization. In R. Lipschutz (Ed.), *On security* (pp. 46–82). Columbia University Press.

World Economic Forum. (2019). *The cybersecurity guide for leaders in today's digital world.* World Economic Forum.

Cybercompliance: A Legal but also Ethical Concept that Allows to Reduce the Current High Risks of Corporations

Bernardino Cortijo Fernández

Abstract The increase in cyber attacks, as well as the mass usage of technology, such as the Internet of Things and M2M connection, together with the millions and millions of personal and organisational data stored in the cloud, put government bodies, organisations and people themselves at constant risk. That is why companies and individuals, given the enormous proliferation of rules, as well as the inclusion of legal entity liability regulations with the intention of reducing risk by forcing the imposition of, in many cases, costly measures, must comply with said rules especially in cyberspace. We call this concept CYBERCOMPLIANCE, and its ultimate goal is to reduce devastating consequences and, in the best-case scenario, eliminate them.

Keywords Compliance · Cybersecurity · Cybercrime · Cyberspace · Cybercompliance

1 Introduction. Cybercrime: A Brief Historical Review

When considering the changes detected in the second decade of the twenty-first century concerning everything that has to do with global security and defence—of nations, localities, companies and citizens—one of the greatest impacts has been CoV2, "probably the worst crisis since the second world war".[i]

Restrictions on movement and business as well as the government's, companies' and citizens' priorities also emphasize existing risks, but also changes in the ways we work and collaborate and the mass usage of collaborative digital media for both professional and personal purposes. On many occasions, users neglect the most basic levels of prevention and precaution, with some even using mass communication systems of a recreational or personal nature for delicate professional matters, exposing documents, information and all kinds of data to any advanced or expert user.

B. C. Fernández (✉)
Nebrija University, Hoyo de Manzanares, Spain
e-mail: bernard.cortijo@dacor-intelligence.com

J. Cayón Peña (ed.), *Security and Defence: Ethical and Legal Challenges in the Face of Current Conflicts*, Advanced Sciences and Technologies for Security Applications,
https://doi.org/10.1007/978-3-030-95939-5_5

This has resulted in technological dependence and an increase in cyberattacks and cybercrimes in general.

The increase in work teleconferences, electronic signatures, smart contracts, trading over the Internet and connected cameras are the basis of the increase in global digitisation (increase of over 55%), mobile devices (over 67%) and Internet accessibility (59%) (Gobierno de España, 2019).

Among the most frequent techniques in cybercrimes are ones related to ransomware extortions, phishing and information theft through attacks or the use of vulnerabilities (EU Agency, 2020; Gobierno de España, 2019).

On the other hand, the need to have solid, auditable and more secure infrastructures that allow one to store millions of data, has increased the usage of cloud or hybrid cloud services (Eurostat, 2021).

Obviously, the risks are evolving towards a situation that makes passive prevention defence very difficult (Crowdstrike, 2020). Furthermore, in the business sphere, protection that does not demonstrate proactivity can, as well be ineffective, become a breach of law. The appearance of crimes or cybercrimes can become a new crime type that is directed at the partners or individuals in charge of the organisation. This criminal offence type can only be reduced with evidence of previous actions that could have been effective, such as training, ethical channels,[ii] the building of a clear risk map and a reviewable mitigation programme and the establishment of control bodies, such as compliance and safety committees.

From all these measures and the need for them to be devised by organisations, private companies, and administrations, in relation to cyber and digital processes and actions, the concept that we define in this chapter as CYBERCOMPLIANCE arises.

What are the factors that influence cyber risks? We have already dealt with the first of them: the cloud. The appearance of the cloud meant an important change in technical processes, in computing, in the definition of systems and applications, but its expansion is still growing, so much so that in the near future it looks like devices will not have to have a local data storage system, nor even an external connection storage device, but will only share the data through cloud fields. This cloud, which is in theory nothing more than backup security, or at least, its delegation in large global storage companies, with all security protocols and policies applied, also represents a concentration of interesting information organised in one place (Knorr, 2020).

IoT (Internet of Things) systems. The possibilities of managing or even being managed remotely or independently, industrial production systems, robotic systems (industry 4.0) and physical equipment with programmed assistance are undoubtedly a clear example of comprehensive digital progress. But at the same time, they represent a possible point of entry into organisational and governmental systems to take control of citizenship and defence systems and activities. This notion has been verified and is not to be overlooked. Any device can be handled or compromised; vehicles, homes, companies, defence or attack equipment, and even in smart city environments, critical elements such as water distribution and storage systems, energy, traffic control, fleets and many others can be attacked.

A more explicit investigation into this matter would focus on M2M (machine to machine), H2M (human to machine), H2H (human to human) or MiH (machine in human) systems, which would undoubtedly allow us to enjoy this concept more (Anton-Haro & Dohler, 2015).

5G and 5G-DIVE. None of this could be done without having communications systems as advanced and still-under-development as the current ones. Steps as clear as 2G, 3G and 4G, ADSL systems and current high-speed fibre systems are predecessors to the now-live 5G.

The use in robot and drone systems has evolved to 5G-DIVE. Speed is useful for development, for new services, for defence and security systems, but also for attack systems and for cyber criminals.

CRYPTOCURRENCY models, current ciphers and BLOCKCHAIN are robust examples in which security is an important issue (McAfee, 2018), but we are already beginning to hear of breaches, some of them based more on a social and human attack than on technological one.[iii]

2 *Modus Operandi* and Trend Evolution in Cybercrime

We are not going to consider a pure technological model nor a theoretical description of the crime types, but rather we are going to make an understandable and clear classification of the reality of these actions, which generally lead to criminal offences.

Society is facing what we could call attacks (MarshMcLennan, 2021). These attacks are on systems (computer, telecommunication, industrial, etc.), privacy and IP (intellectual or industrial property), and in many cases involve data or information 'theft'.

Attacks that are on the rise in the post-pandemic era include phishing techniques, identity theft, internal threats, information leakage or theft and ransomware (Gobierno de Espana, 2019).

Others are still very active, such as the use of malware, website and application attacks, all kinds of data security breaches, physical damage, theft and loss of devices and equipment (Sonicwall, 2021; TrendMicro, 2021).

Spam, DDoS or distributed denial-of-service attacks (probably due to the security systems used), the use of BOTNETS, cyber espionage and cryptojacking have decreased slightly, although they still occur frequently (FBI, 2021).

3 Cyber Objectives and Regulations

3.1 Cyber Objectives

Pre-COVID, COVID and post-COVID times have witnessed changes[iv]—not radical, but notable—and organisations connected in one way or another to COVID or health care have been an important target (UN, 2020). Previously, patients and healthcare settings, and therefore, healthcare centers, hospitals and clinics, were seen as a point of search for information and attacks, mainly on privacy and data access.

Banks and financial groups, along with the rest of the critical infrastructures such as telecommunications, industry (the advance of autonomous systems and 4.0 have a significant influence) are also a target.[v]

Public institutions and governments continue to be of great interest, especially during election periods or crises, as well as educational centres (universities and schools), the former for activism or influence, and the latter for data collection.

Finally, we would highlight service and retail companies, as well as an increase in interest in small- and medium-sized companies, as well as personal data, mainly related to applications and services.

3.2 Regulations

In relation to the prolific existing regulations, we have a large number of concrete samples, although they are not global.

In Spain, for example, there are regulations in force related to national security that include partial or total aspects of cybersecurity and cybercrime:

Ley de Seguridad Nacional
Consejo Nacional de Seguridad
Estrategia de Seguridad Nacional
Esquema Nacional de seguridad en el ámbito de la Administración Electrónica
Comité de Seguridad del Sistema de Información de IT
Comité de Seguridad de la Seguridad Social
Seguridad de Redes y Sistemas de Información
Ley sobre Secretos Oficiales
Ley de Secretos Empresariales
Estrategia Nacional de Ciberseguridad.

Regulations related to Critical Infrastructures, which the European Union has conditioned to carry out:

Ley y Reglamento de medidas para la protección de las Infraestructuras Críticas
Public and private security regulations:
Parcial de Desarrollo de la Dirección General de la Policía

Ley de Seguridad Ciudadana
Ley de Seguridad Privada.

About security incident response teams:

Ley de Servicios de la Sociedad de la Información y de Comercio Electrónico
Centro Criptológico Nacional
Normativas para el ejército y defensa.

About Telecommunications:

Ley de Servicios de Seguridad de la Información y Comercio Electrónico
Distintivo Público de Confianza en los servicios de la sociedad de la información
Adecuación y funcionamiento del sector público por medios electrónicos
Servicios electrónicos de confianza
Certificados de firma electrónica y documento nacional de identidad
Ley General de Telecomunicaciones
Uso de dominio público del espacio radioeléctrico
Conservación de datos en comunicaciones electrónicas
Entrega de datos por los operadores de telecomunicaciones.

Especially in Cybercrime:

Ley del Código Penal
Ley de Enjuiciamiento Criminal (procesal)
Modificaciones sobre la responsabilidad penal de los menores.

About data protection:

Ley de Protección de datos personales y garantías de los derechos digitales
Reglamento Europeo
Ley de protección de datos para juicios penales.

And this is just a Spanish national panorama. At the European level, let us cite a few:

Budapest Cybercrime Convention
Directive 2016/1148 on security measures in networks and information systems of the European Union
Regulation on Protection of Personal Data
E-evidence directive.

There is a lot of directives, but there are no global international agreements.

Russia made a proposal for a new Convention on Cybercrime in 2010 that was clearly reject by the UN. It seems that the trend to coordinate investigations regardless of legislation is taking shape. This was partly considered in the US Cloud Act and the proposed e-Evidence Directive.

A working group. Has been crated in the UN, in the environment of organized crime, for the study of Cybercrime, given the little usefulness of the Budapest Convention.

EU Cybersecurity Strategy
EU Cybersecurity Regulation
European cybersecurity certification system
Directive on networks and systems
Directive on attacks on information systems
Resolution on encryption.

Although it is true that Spanish regulations are among the most extensive in the world, in recent years we have a great legislative capacity in a large part of countries outside the European Union as well, and clearly reflects a rend of regulation more based on covering needs and specific aspects to address the problem globally and internationally.

However, there are rules and also internal regulations, as we will see, so the ability to verify compliance, audits and reviews (Compliance) both in the criminal aspect and in the more technology-oriented regulations takes more strength.

4 Sequence of Conclusions

4.1 Regulations, Too Many or Not Enough? Problems

According to the seen, we can say and believe that there is a large number of criminal and civil rules that deal with this problem. Part of the difficulty is the various existing rules, with different criteria, with different and complex criminal processes, and what is even worse, with international or extra-national regulations that prevent the resolution of investigations, the clarification of crimes and the execution of justice, in an environment that in itself is supranational.

But the reality is that there are rules, maybe not enough, maybe not in the right jurisdictions, maybe not technology-adjusted, but there are rules.

4.2 Internal Policies and Rules

On the other hand, organisations and institutions themselves, outside of the activities of House lawmakers, must impose their own criteria and responsibility to put global, security, control and supervision policies in place, as well as specific regulations for specific processes and activities. This self-regulation may go beyond what is required by law, but not what is required for society. An example of this are proposals to support minors, disadvantaged groups, natural resources and, ultimately, the planet itself.

4.3 Cybercompliance. Legal Entity Liability. An Analysis Focused on Compliance in Cyberspace

We'll finish with compliance activities and, more specifically, activities related to systems and processes in cyberspace and digital environments, which we will call cybercompliance, as we agreed at the beginning.

We must not only process risk mitigations detected in compliance risk analyses (mainly criminal and reputational), but also measures for early detection of hacking actions and cyber attacks, detection of data theft and personal data protection monitoring, systems for detecting data and system attacks, interceptions or violations of communications, information classification systems and the implementation of their protection, analyses of individuals involved in business (with the help of cyber intelligence systems), and in general, meticulous organisational and technological control, regular measures and audits that, in general, are drawn up in a Compliance Management System.

Everything is encompassed in a process of organisational internal responsibility, public or private, company or institution, that goes beyond the strict compliance with rules—as there are crimes such as legal entity liability and company liability—and which, far from being a mere formality, is a matter of practical defence consequences.

The increase and variety of cyberattacks, the use of massive cloud storage, and IoT systems put organizations and their users or customers at risk. So diverse norms, and sometimes many and scattered, are hardly usable, in a multi-geography. Compliance requires a description of the risks existing in each organization and the development of a risk map. It also offers the ability without retaliation to employees to alert of possible illegal or unethical actions. It also forces awareness of a culture of compliance and knowledge of standards and technology along with risks, reducing their appearance. Therefore, internal compliance verification and criminal, technology and ethical audit routines will reduce these risks.

Notes

(i) Antonio Guterres, UN Secretary General in his telematic appearance from UN headquarters in April 2020.

(ii) Whistleblowing Directive already incorporated into the EU (Directive UE 2019/1937 European Parliament. October 2019).

(iii) "private blockchain users are often the easiest target, due to start up mentality, in which security is relegated to the background" (McAfee, 2018).

(iv) Interesting to analyze the document made by Marsh "The Changing Face of Cyber Claims 2021".

(v) Fintech's, Cooperatives, Socaps, Sofipos includes. And electricity, water supply, transportation, security and emergences services.

References

Anton-Haro, C., & Dohler, M. (2015). *Machine-to-machine (M2M) communications: Architecture*. Woodhead Publishing.

Crowdstrike. (2020, September 22). *Threat hunting report*. Retrieved from https://www.crowdstrike.com/resources/reports/threat-hunting-report-2020-es/

EU Agency. European Union Agency for Cybersecurity. (2020). *COVID19: Stronger together in fighting cyber threats*. Retrieved from https://www.enisa.europa.eu/topics/wfh-covid19/media/covid19-stronger-together-in-fighting-cyber-threats

Eurostat. (2021, January). *Cloud computing—Statistics on the use by enterprises*. Retrieved from https://ec.europa.eu/eurostat/statistics-explained/index.php?title=Cloud_computing_-_statistics_on_the_use_by_enterprises

Federal Bureau of Investigation, Internet Crime Complaint Center. (2021, March). *Internet crime report 2020*. Retrieved from https://www.ic3.gov/Media/PDF/AnnualReport/2020_IC3Report.pdf

Gobierno de España. (2019). *Informe Anual de Seguridad Nacional*. Retrieved from https://www.dsn.gob.es/es/documento/informe-anual-seguridad-nacional-2019

Knorr, E. (2020, June 8). *The 2020 IDG cloud computing survey*. Retrieved from https://www.infoworld.com/article/3561269/the-2020-idg-cloud-computing-survey.html

MarshMcLennan. (2021). *The changing face of cyber claims 2021*. Retrieved from https://www.marsh.com/es/es/services/cyber-risk/insights/the-changing-face-of-cyber-claims-2021.html

McAfee. (2018). *Informe sobre amenazas contra blockchain*. Retrieved from https://www.mcafee.com/enterprise/es-es/assets/reports/rp-blockchain-security-risks.pdf

Sonicwall. (2021). *Cyber threat report: Cyber threat intelligence for navigating the new business reality*. Retrieved from https://www.sonicwall.com/2021-cyber-threat-report/

Trend Micro. (2021, September 14). *Attack from all angles: Trend micro 2021 midyear cybersecurity report*. Retrieved from https://www.trendmicro.com/vinfo/us/security/research-and-analysis/threat-reports/roundup/attacks-from-all-angles-2021-midyear-security-roundup

United Nations. (2020, September). *UN comprehensive response to COVID-19*. Retrieved from https://www.un.org/sites/un2.un.org/files/un-comprehensive-response-to-covid-19.pdf

Criminal Law and the Risks Posed by Internet. An Examination of Spanish Criminal Law

Pilar Otero

Abstract The vertiginous advance of technology has seen the advent of new forms of crime such as hacking, phishing, child grooming, cracking and denial of service, and also the use of computer networks to commit traditional offences such as fraud, harassment, defamation, intimidation, industrial espionage, and child pornography. It is fair to say that today's society has become a society of risk. Cyberspace has become a haven for criminality and a platform for organised crime. The specific characteristics of information and communication technologies (ICTs) require a new response from the field of criminal law in order to tackle cybercrime. Thus, it is necessary in some situations *to advance the punitive barrier* (to use the Spanish expression), that is, to bring forward the moment when the conduct is considered an offence; in others, to reformulate certain criminal concepts in order to adapt to the new scenario; and in still others, to create new criminal offences (though related to existing ones) since the new types of criminal activity do not conform to the definitions traditionally used in criminal law. Furthermore, the cross-border nature of these activities means that an efficient response of the system requires the reinforcement of international cooperation mechanisms; the harmonisation of criminal provisions; and the enhancement of effective procedural instruments, even though this may imply a certain renunciation of sovereignty on the part of the States concerned. This is the main purpose of the creation of supranational regulations, represented above all by the Council of Europe Convention on cybercrime (Budapest, 23.11.2001) and Directive 2013/40/EU of the European Parliament and of the Council on attacks against information systems.

Keywords Cybercrime · Hacking · Cracking · Phishing · European directive 2013/40/EU · Criminal law · Transnationality

P. Otero (✉)
Carlos III University, Getafe, Spain
e-mail: potero@der-pu.uc3m.es

J. Cayón Peña (ed.), *Security and Defence: Ethical and Legal Challenges in the Face of Current Conflicts*, Advanced Sciences and Technologies for Security Applications, https://doi.org/10.1007/978-3-030-95939-5_6

1 The Background

The vertiginous technological advances of recent decades have radically transformed our lives, invading almost all areas of human activity. They have also generated what has been termed a *society of risk*. Among the manifestations of the dark side of this technological development we might mention the following:

1. New forms of crime. The crime of hacking, or illegal access to an information system (art. 197 *bis*, Spanish penal code, and art. 323-1, French penal code), the crime of phishing or computer fraud (art. 248.2, Spanish penal code; §263ª, German penal code, henceforth StGB), the crime of child grooming or cyberbullying of minors (art. 183 *ter*, Spanish penal code), the crime of cracking, i.e., cyber damage or computer sabotage (arts. 264 ff., Spanish penal code), and denial of service attacks, i.e., the disabling of an information system (art. 264 *bis*, Spanish penal code and art. 617 *quater*, Italian penal code).
2. The use of computer networks to commit traditional offences such as fraud, harassment, defamation, intimidation, industrial espionage, child pornography, credit card cloning, and so on.

   Therefore, the term "cybercrime" may refer either to offences that can only be committed via Internet (for instance, hacking or computer sabotage, i.e., attacks on the information system itself) or to other traditional offences that use telematic networks as a means to achieve their ends, thus violating various legal rights such as privacy, property, etc. (Álvarez Vizcaya, 2001; Morales García, 2010).
3. The spread of the effect of a criminal act to an unlimited number of users at breakneck speed: for example, the victimisation framework generated by computer viruses.
4. The transnational nature of offences, which makes their prosecution difficult (Flores Prada, 2012). The absence of borders is an inherent feature of the Internet: the Web is a communication infrastructure, a network of networks interconnecting innumerable groups of computers that bring together millions of people with no constraints of time or distance. Not only does the Internet know no borders, but it is not controlled by any entity or institution. There is no central public authority governing the Internet; no one is in charge (López Ortega, 2001). The upshot is that private entities, acting in free competition and in accordance with market rules, establish themselves as monopolies and have no difficulty in evading the authority of individual States (Llaneza, 2000).
5. The difficulty of sanctioning natural or legal persons who provide services on the Internet (service providers) for failing to remove criminal contents. Their criminal liability is based on two criteria, both of which are difficult to demonstrate (Morales García, 2001): prior knowledge of the existence of illegality, and diligence in proceeding to prevent access, set out in Directive 2000/31/EC of the European Parliament and of the Council, of 8 June, 2000, which foresees a system of horizontal accountability, common to all cases of illegal content: child pornography, defamation, and incitement to racial violence and which was

introduced into the Spanish legislation by Law 34/2002, of 11 July, *on services of the information society and electronic commerce.*

6. The dissociation in terms of both place and time between the offence and its outcome. This may make it particularly difficult to determine responsibility for, or participation in, the crime. More and more frequently, the offender does not need to be present in the place of the crime (Lezertúa, 2001). This situation vastly increases the difficulty of prosecution; in countries like the US, it is believed that only 1% of computer-related crimes committed are detected; of these, only 14% are brought to trial, and only 3% of the cases heard result in imprisonment.
7. The greater the complexity of the system, the greater its vulnerability.

All these points can be summed up in one essential idea: the Internet favours anonymity. Although IP addresses can now be identified more easily, and although cybercrime leaves digital traces, the truth is that it remains more difficult to identify perpetrators of these virtual offences than perpetrators of similar offences in the real world. The perception of anonymity inevitably enhances the feeling of impunity among cybercriminals and, as a result, means that they are more likely to reoffend (Miró Llinares 2012, 2016–2017). Cyberspace thus becomes a haven for criminality and an ideal platform for organised crime.

Furthermore, the Internet is the conduit through which most of the world's money circulates. This means that cybercrime mainly has an economic purpose. Any cybercrime, even if its immediate effect is on other legal assets such as privacy or the security of information systems, ultimately has a fundamentally economic interest. Child pornography, for example, an illegal activity which violates the idea of minors' freedom from sexual abuse, currently moves the chilling figure of a billion euros a month.[i]

2 The Response of Criminal Law

How has the Spanish criminal law system met these challenges? Law, in general, and criminal law, in particular, is always one step behind the new communication and information technologies. In the area that concerns us here, this situation has had several important consequences:

2.1 The Advancement of Punitive Barriers: The Application of the Concept of Crimes of Abstract Danger

The response of criminal law in Spain shows signs of disorientation, manifested fundamentally in the increase in the application of the concept of *crimes of abstract danger*: that is, offences in which an effective situation of danger is not produced, but are sanctionable on the grounds that they have the potential to produce such a situation. Thus, an instance in which the legal asset in question has not yet been violated may qualify as an offence (Galán Muñoz, 2009). In addition to creating a serious degree of legal uncertainty, this also leads to a weakening of legal guarantees. In this scenario, a person who manufactures or possesses computer programs specifically designed to commit an Internet fraud receives the same punishment as the person who actually carries out the fraud (art. 248 a and b, Spanish penal code).

2.2 The Reformulation of Certain Criminal Concepts

In order to adapt to this new situation, Spanish criminal law has been forced to develop its conception of certain criminal categories. The two examples may be illustrative.

First, the concept of *criminal damage* in the real world is characterised by the enduring effect of the damage, and its classification is linked to its economic value. That is, if the economic value of the damage committed is above € 400, the offence is considered serious; below this figure, it is considered minor.

In the virtual world, however, for *cyber damage* to be considered an offence, it must be *serious*; defined as such not necessarily in terms of its economic value (how much does a USB cost: € 5–10?), but in terms of the functional value of the data contained in the program affected (Fernández Teruelo, 2007; Andrés Domínguez; Morales García, 2010). This interpretation assumes criteria traditionally linked to civil liability, such as the cost of cleaning the computer system, whether or not there was a backup copy, the amount of information lost, the hours of work taken to prepare the document, and so on. The enduring effect, or permanence, is not necessarily a criterion for cyber damage: for instance, the offence of denial of service may involve obstructing or interrupting the operation of a computer system without destroying or damaging it (for example, by bombarding the system with spam messages and thus rendering it inoperative).

The second example is the evolution of the concept of *intimacy* in Spanish criminal law into a criminally protectable legal asset in the virtual world. Previously understood as a prohibition of external interference, with the expansion of communication technologies such as mobile telephones in the 1980s, *intimacy* has now been replaced by the concept of *privacy* characteristic of the English-speaking world (Morales Prats, 2002), It is thus conceived as an active right of control, linked to the idea of the individual's self-determination. *Privacy* thus refers to a set of characteristics of the person which in isolation lack significance, but which together make up a digital portrait of the individual's personality. However, with the emergence of the Internet, users leave traces or fingerprints; network access providers record the time and location of connections (in log files); cookies (files emitted by the web servers visited by users, and which are recorded on their hard drives) maintain users' browsing histories. Given the difficulty of maintaining privacy on the Internet, the concept has been redefined as the *right to anonymity*. However, for the reasons mentioned above, this *right to anonymity* cannot be fully enforced; this is why it is currently limited to the *right to be forgotten*. In this regard, the Sentence of the Court of Justice of the European Union (CJEU 13-5-2014, case c-131/12) which transfers the responsibility for deleting data to the search manager, is especially interesting, although CJEU 24-9-2019 limited its scope to the EU itself.

2.3 The Creation of New Offences

As we have seen, the nature of ICTs means that they cannot always be regulated through the application of traditional criminal offences. A new response is required, for example, to deal with phishing, the most widespread form of Internet fraud. Phishing is the manipulation of a computer system to achieve a non-consensual transfer of property. This activity diverges slightly from the conventional notion of fraud; the defining feature of fraud is the deception of the victim, who, as a result, voluntarily carries out an action that is contrary to his/her interests. Note that in the case of phishing, the perpetrator of the manipulation of the computer system is also the perpetrator of the money transfer, without the victim realising; so this is a case of telematic theft rather than of fraud. Due to the requirements of the principle of criminal legality, given the impossibility of classifying these behaviours under the heading of fraud, a reformulation had to be introduced that substitutes *deception* with *manipulation*.

Let us look at several examples. Traditional fraud is typified in art. 248 of the Spanish penal code, while phishing is regulated in art. 248.2 a). Much the same happens in other legal-criminal systems. For instance, fraud is included in §263 of the German StGB and computer fraud in §263[a], which removes the concept of deception and error present in the traditional notion of fraud and places the emphasis on the loss of property suffered by the victim. Similarly, the Italian penal code, which defines fraud in art. 640 and computer fraud in art. 640 *ter*, also removes the notions of deception and error, and prioritises the procurement of an unfair benefit via the alteration of the functioning of a computer system. In fact, the Italian penal code had to create a new offence to cover cases of damage to software or to data, since it was impossible to define these events as *physical objects*, which are covered in art. 635 of the code.

3 Transborder Issues; the Difficulties of Prosecution

Given that cybercrime is an activity that crosses national borders, an efficient response of the system requires the reinforcement of international cooperation mechanisms, the harmonisation of legal provisions and the development of effective procedural instruments, even though in certain areas this may imply a certain surrender of sovereignty by the States. This is the main purpose of the supranational regulations in this regard, represented primarily by the Council of Europe Convention on cybercrime (Budapest, 11-23-2001), the Council's Framework Decision 2005/222/JHA, on attacks against information systems and Directive 2013/40/EU of the European Parliament and of the Council, on attacks against information systems, which replaces the earlier Framework Decision.

3.1 The Harmonisation of Legal Provisions

In compliance with the Framework Decision of 2005, the following criminal offences were included for the first time in the Spanish penal code (by means of Organic Law 5/2010): (a) illegal access to information systems (hacking) and (b) illegal interference in information systems (cyber damage, computer sabotage or cracking).

Most European legal-criminal systems did not find an appropriate place for these offences, since European regulations (Framework Decision 2005/222/JAI of the Council and Directive 2013/40/EU of the European Parliament and of the Council) laid down that they should be included in an independent chapter in which the legal asset protected is the security of the information systems. The crime of hacking was inadequately introduced into the Spanish penal code since it was interpreted not as a crime against network security, but as a violation of privacy (Carrasco, 2010). This has caused a significant number of problems of interpretation, as will be specified later. As for cracking, it was also incorrectly included as a crime against property, when European regulations state that it should be incorporated in reference to the security of information systems. Indeed, as noted above, an information system may be disabled without its data being definitively destroyed; and the quantification of the damage may represent a further difficulty.

This same form of transposition has been followed by other countries such as Germany, which places data integrity and theft in §§ 202^{a}-202d of the StGB (under the violation of privacy) while the alteration of data and computer sabotage is found in 303a and 303 (under damage to property). The Italian penal code, for its part, includes hacking in art. 615 *ter*, under crimes against the inviolability of the home, as if it were a cyber burglary (i.e., unauthorised access to a computer or telematic system). Computer sabotage in the Italian penal code is also classified in art. 635 *bis* (under crimes against property).

Other countries in our vicinity, however, have applied the European Directive correctly. The French penal code, for example, places cybercrime in the chapter *Attacks on automated data processing systems*, including both illegal access to the information system and the destruction, obstruction or modification of data in arts. 323-1 to 323-8, and thus protects a single, unified legal asset.

Directive 2013/40/EU has been transposed into the Spanish penal code (by Organic Law 1/2015) and into other European penal codes (in Italy, France and Germany). This has expanded the scope of the offences previously defined, and has also led to the introduction of new ones:

3.1.1 In Relation to the Illegal Access to Information Systems

1. Accessing, or facilitating access to, a *computer system or part of it*, violating the security measures established to prevent this, or remaining in the system against the will of whoever has the legitimate right to exclude this, is considered an offence (art. 197 *bis*, Spanish penal code). Prior to the reform, the crime of hacking required the accessing of the *data* contained in the information system; now, it is enough to access the *information system* or part of it. This means, first, that the concept of a *crime of abstract danger* is applied; and second, that even though it is still classified under violations of privacy, since access to the system is sufficient for the crime to be committed the orientation of the protection now moves towards the *security of the information system* (access to the *data contained in the system*, which was required before the reform, is a more obvious violation of *privacy*).
2. The interception of data transmitted between computers on the network—that is, between systems—is also included (art. 197 *bis* 2, Spanish penal code).
3. In accordance with the policy of punishing preparatory acts, not only those persons who access but those who, without due authorisation, produce, acquire for their use, import or in any way provide third parties with a computer program conceived or otherwise adapted principally to commit the crime of hacking, or a computer password, an access code or similar data that allow access to all or part of an information system, are liable to punishment (see art. 197 *ter* of the Spanish penal code, art. 615 *quater* of the Italian penal code which sanctions the mere unauthorised possession of access codes to computer or telematic systems; or art. 323-3-1 of the French penal code, which sanctions the retention, offering or making available of a computer program especially adapted to commit the crime of hacking).
4. Legal persons who commit these offences are to be held criminally responsible (art. 197 *quinquies* of the Spanish penal code; 323-6 of the French penal code). The offences are considered to be more serious when carried out within an organisation or criminal group (art. 197 *quater*, Spanish penal code or 323-4 and 323-4-1, French penal code).
5. Finally, hacking that involves the unauthorised use of the victim's personal data is deemed particularly serious. It seems that the Spanish legislators intended to transpose Directive 2013/40/EU, which establishes this obligation on the Member States. However, this requirement of aggravating circumstances was not introduced in the appropriate place in the penal code; it does not refer to attacks on information systems, and the indications of the Directive that justify harsher sentencing, such as the purpose of gaining the trust of a third party or causing damage to the legitimate owner of the identity, have not been added. Thus, the interpretation of identity theft required by the Directive has not been included in the Spanish regulations, and so there are difficulties of delimitation with regard to other criminal offences that affect privacy (Otero, 2016–2017).

The Circular of the State Attorney General's Office (FGE) 3/2017[1] adds that identify theft refers not only to official identity data but also to any other data that identify an individual or make it possible to identify him/her in front of third parties, in either a physical or a virtual environment.

3.1.2 In Relation to Crimes of Cyber Damage

1. The offence of the abuse of devices created to produce computer sabotage is defined in art. 264 *ter*. Similarly, the installation of instruments capable of intercepting or interrupting a computer system is punishable, even if the system is not actually interrupted (art. 617 *quinquies* of the Italian penal code; 323-3-1 of the French penal code).
2. The express inclusion of DoS (Denial of Service) attacks is maintained as a specific form of cyber damage (art. 264 *bis* of the Spanish penal code; 617 *quater* of the Italian penal code; 323-2 French penal code).

It should be noted that cyber damage includes the destruction or disabling both of complete computer systems and of their specific components: data, documents or programs (De La Mata, 2016).

3. Aggravating circumstances are considered when cyber damage targets key infrastructures, the disruption of which may have significant cross-border repercussions. See, for example, art. 264.1.4 of the Spanish penal code or "crimes of attack" against data and computer systems of public utility (art. 635 *ter*, paragraph 1 and art. 635 *quinquies*, paragraph 1, of the Italian penal code), or when these crimes are committed within a criminal organisation (art. 264.1.1, Spanish penal code).

The fact that Spain has not directly safeguarded the security of information systems, but rather protects, albeit remotely, privacy (against hacking) or property (against cyber damage), raises problems of competing jurisdictions and may also present a violation of the principle *non bis in idem*, especially with regard to hacking and its relationship to the offences defined in arts. 197 et seq., which protect other aspects of privacy such as emails or data of a personal nature recorded in files or other digital, electronic or telematic means. In this regard, the Circular of the General State Attorney's Office 3/2017, as well as the case law (see the Supreme Court Sentence 24-2-2015) have provided various solutions in case of conflict. Thus, hacking should be interpreted as access to data or information that may affect privacy but are not directly related to personal *intimacy*, in the Spanish legal sense: having access to the personal list of contacts is not the same as compiling data related to the version of software used or the location of a system's input ports. For this reason, a separate, distinct definition of the mere access to computer systems is preferred. Consequently, the precept that regulates hacking applies unless the privacy of the individual is

[1] https://www.boe.es/buscar/abrir_fiscalia.php?id=FIS-C-2017-00003.pdf.

affected – in which case arts. 197. 1 and 2 of the Spanish penal code are applicable (which, furthermore, entail a harsher penalty).

3.2 Procedural Instruments

The harmonisation of legal provisions is of no use if there are no effective procedural instruments to prosecute the illicit behaviours to which they refer. The very nature of these behaviours, characterised by the use of sophisticated technical procedures, complicates the prosecution of crimes. Indeed, their prosecution requires the use of ICTs themselves as an investigative tool, with the support of specialists such as cyber undercover agents.

The main difficulty facing investigators in this area lies in finding a balance between investigative efficiency, i.e., overcoming the technical obstacles that surround these crimes, and the strict observance of the legal rights of the persons under investigation. That is to say, the electronic evidence (namely, the information generated, stored, or transmitted through the use of electronic devices capable of obtaining a judicial conviction) is easy to manipulate or destroy. On the other hand, the rights of the person under investigation must be upheld; this is true of all criminal proceedings, but is particularly important in the investigation of technology-based crime, because these proceedings affect fundamental rights such as privacy or the confidentiality of communications.

This counterweight is being achieved in Spain based on Organic Law 13/2015, 5-10, *strengthening the procedural guarantees and the regulation of technological research measures*. Additionally, we cannot ignore the doctrine of the Court of Justice of the European Union in this area which, as a result of the relevant judgment on electronic communications of 8-4-2014 (joined cases C 293/12 and C 594/12), the recent sentence of the CJEU of 6-10-2020 (joined cases C 511/18, C 512/18 and C 520/18 *La Quadrature du Net* and others) and the sentence of the CJEU 2-3-2021 (Prokurator case) have maintained that a generalised and unselective preservation of related traffic data[ii] and location data cannot be established on a preventive basis. In any process of communication, in addition to the content communicated, other data are generated, and their tracking leads to the identification of the work station where the communication originated. However, only the fight against *serious crime that threatens national security* can justify the preservation of these data by providers of access to online communication services or by providers of storage services. In addition, to be legal, this preservation must be subject to the principles of speciality, suitability, exceptionality, necessity, proportionality of the limiting measure, temporality, and judicial control.

3.3 Areas of Competence and International Cooperation Between Police and Legal Systems

We now turn to the question of area of jurisdiction: that is, who has the competence to prosecute these crimes. Given their transnational nature, emphasis should be placed on developing a common global interpretation of the principle of ubiquity (according to which a crime may be prosecuted in all jurisdictions in which some part of the offence has been carried out; consequently, investigating judges in any of these jurisdictions will in principle have powers to prosecute the case).

As a complement to the above, the coordination and collaboration between States is essential: on the one hand, to ensure an optimal determination of jurisdiction and to set out clearly the criteria of prioritisation in case of conflict; and, on the other, to strengthen instruments of international law enforcement and cooperation such as Europol and Interpol.

Thus, art. 12 of Directive 2013/40/EU determines the following with respect to jurisdiction:

1. Member States shall establish their jurisdiction with regard to the offences referred to in Arts. 3 to 8 where the offence has been committed:
(a) in whole or in part within their territory; or
(b) by one of their nationals, at least in cases where the act is an offence where it was committed.
2. When establishing jurisdiction in accordance with point (a) of paragraph 1, a Member State shall ensure that it has jurisdiction where:
(a) the offender commits the offence when physically present on its territory, whether or not the offence is against an information system on its territory; or
(b) the offence is against an information system on its territory, whether or not the offender commits the offence when physically present on its territory.
3. A Member State shall inform the Commission where it decides to establish jurisdiction over an offence referred to in Arts. 3 to 8 committed outside its territory, including where:
(a) the offender has his or her habitual residence in its territory; or
(b) the offence is committed for the benefit of a legal person established in its territory.

4 Conclusions

Criminal law is advancing steadily, but slowly, in its attempts to contain the rampant phenomenon of cybercrime. Whenever it manages to offer a solution to a problem, another new form of criminal activity crops up. In times of crisis such as the Covid-19 pandemic, new computer frauds have proliferated—for example, those that use increasingly sophisticated forms of social engineering. These developments represent a new challenge for the legal instruments currently in place.

The transnational nature of cybercrime has shown that in order to find effective solutions to the *modus operandi* of the perpetrators it is essential, first of all, to harmonise the legal provisions of different States. Given that this crime knows no borders, there is no point in defining the crime of hacking in Spain if it is not defined in other countries in our vicinity. Secondly, a balance must be found between procedural instruments that use specialised investigating tools or personnel to compile electronic evidence (for example, cyber undercover agents) and the need to respect the rights of the person under investigation, since the procedures carried out in the investigation of technological crime impinge on fundamental rights such as privacy and the confidentiality of communications.

Coordination and collaboration between States is essential in order to ensure an optimal determination of jurisdiction and to clearly set out the criteria for prioritisation in case of conflict, and also to strengthen the instruments of international police cooperation such as Europol and Interpol.

All of this is being achieved thanks to European regulations, represented mainly by Directive 2013/40/EU—which envisages the transposition of its provisions into the internal regulations of the Member States—and the doctrine of the Court of Justice of the European Union, which lays down the guidelines on the preservation of data deriving from electronic communications.

Notes

(i) Cfr.https://www.usc.gal/export9/sites/webinstitucional/gl/institutos/criminoloxia/descargas/Pornografxa_Infantil.pdf

(ii) *Related traffic data* are data generated as a consequence of a communication via an electronic communications network, of its being made available to the user, or of the provision of an ICT service of a similar nature (art. 588 *ter* b, LO 13/2015, 5–10).

Acknowledgements This study was carried out as part of the research project "H2019/HUM-5699 (ON TRUST-CM. Interuniversity program in culture of legality), financed by the Community of Madrid and the European Social Fund" and the research projet "Criminal responsibility of transnational corporations for violations of human rights and the environment" (DER2017-85144-C2).

References

Álvarez Vizcaya, M. (2001). Consideraciones político criminales sobre la delincuencia informática: el papel del Derecho penal en la Red. *Internet y Derecho penal*, Madrid, Consejo General del Poder Judicial, pp. 255 y ss.

Andrés Domínguez, Ana Cristina. (2010). Daños. *Comentarios a la reforma penal de 2010*, Valencia, Tirant lo Blanch, pp. 291–296.

Carrasco Andrino. (2010, Mar), "El delito de acceso ilícito a los sistemas informáticos. *Comentarios a la reforma penal de 2010*, Valencia Tirant lo Blanch, pp. 249–256.

Circular Fiscalía General Del ESTADO 3/2017, 21-9.
Code Pénal Française, modifié par LOI nº 2015-912 du 24 juillet 2015-art- 4.
Codice Penale Italiano D.L. nº 130/2020, 21 ottobre 2020 con modificazioni dalla L. 18 diciembre 2020, n. 173.
Codigo Penal Español (LO 10/1995, 23-11) modificado por LO 1/2015, 30-3.
De La Mata Barranco, Norberto (2016). Los delitos contra la integridad y disponibilidad de datos y sistemas informáticos después de la LO 1/2015", *Estudios de Derecho penal. Homenaje al profesor Miguel Bajo*, Madrid, Ramón Areces, pp. 1089–1108.
Fernández Teruelo, J. (2007). *Cibercrimen. Los delitos cometidos a través de Internet*, Oviedo, Constitutio Criminalis Carolina.
Flores Prada, I. (2012). *Criminalidad informática. Aspectos sustantivos y procesales*, Valencia, Tirant lo Blanch.
Galán Muñoz, A. (2009). *Libertad de expresión y responsabilidad penal por contenidos ajenos en Internet*, Valencia, Tirant lo Blanch.
González, P. O. (2016–2017). Delitos contra la intimidad, derecho a la propia imagen e inviolabilidad del domicilio. *Memento práctico penal económico y de la empresa*, Madrid: Francis Lefebvre.
Cibercrímenes económicos y patrimoniales. (2016–2017). *Memento práctico penal económico y de la empresa*, Madrid: Francis Lefebvre.
Lezertúa Rodríguez, M. (2001). El Proyecto de Convenio sobre el cybercrimen del Consejo de Europa. *Internet y Derecho Penal*, Madrid: Consejo General del Poder Judicial. Servicio de formación continua, Escuela Judicial.
Llaneza González, P. (2000). *Internet y comunicaciones digitales*, Barcelona: Bosch.
López Ortega, J. J. (2001). Libertad de expresión y responsabilidad por los contenidos en la Red. *Internet y Derecho Penal*, Madrid: Consejo General del Poder Judicial. Servicio de formación continua, Escuela Judicial
Miró Llinares, F. (2012). *El cibercrimen fenomenología y criminología de la delincuencia en el ciberespacio*. Madrid: Marcial Pons.
Morales Prats, F. (2002). Internet: riesgos para la intimidad. *Internet y Derecho penal*, Madrid, Cuadernos de Derecho Judicial.
Morales García, O. (2001). Criterios de atribución de responsabilidad penal a los prestadores de servicios e intermediarios de la Sociedad de la Información", *Revista de Derecho y Proceso Penal*, nº 5, Aranzadi, pp. 139 a 167.
Morales García, O. (2010) Delincuencia informática: intrusismo, sabotaje informático y uso ilícito de tarjetas. *La reforma penal de 2010: análisis y comentarios*, Pamplona, Aranzadi, pp. 181–194.
StGB (Strafgesetzbuch) Deutschland. (2015, Dezember). Bundesgesetzblatt (BGBI) I. S. 2218, geändert worden ist.

Human Rights, Big Data and Artificial Intelligence: Elements of a Complex Algorithm

María Méndez Rocasolano

Abstract As individuals, we are under the make-up power of new technology used by other powers (multinationals, banks, politics-ideologies) because machines have an enormous computing and processing power of information that can be used to direct and control wills. Through the study of big data, and artificial intelligence legal concepts, and their regulation, we can propose juridical considerations as the use of the *Dynamogenesis of values and law* linked with the Design for Values, HRESIA (Human Rights, Ethical and Social Impact Assessment) model to create spaces of human security and freedom in a future where the power of technology could become one of the state powers. In this paper, drawing on methodologies of Value Sensitive Design and Participatory Design to present a roadmap for proactively engaging societal stakeholders to translate fundamental human rights into context-dependent design requirements through a structured, inclusive, and transparent process.

Keyword Personal data protection · Human rights · Big data · Artificial intelligence dynamogenesis of values · Human dignity

1 Consequences of the Big Data and Artificial Intelligence in the Digital Society

The concept of data is not new, and its use is not necessarily a human characteristic. However, the concept is currently found in a specific context, where it appears inextricably linked to three phenomena, digitalisation, microdata and Internet. The interaction of these three elements has changed our perception of reality. In some manner, we have amplified both our capacity to perceive information about the reality that surrounds us, and our capacity to process and convey information, and interact

This article has been developed within the research group: Law, Ecotechnology and Innovation: keys to 21st Century Development. GI-UCAM-DEI4SXXI.

M. M. Rocasolano (✉)
UCAM Catholic University of Murcia, Murcia, Spain
e-mail: mmrocasolano@ucam.edu

J. Cayón Peña (ed.), *Security and Defence: Ethical and Legal Challenges in the Face of Current Conflicts*, Advanced Sciences and Technologies for Security Applications,
https://doi.org/10.1007/978-3-030-95939-5_7

with other amplified human beings. At the same time, we are living in a unprecedented rate expanding virtual world. In fact, it exceeds human intervention, creating an autonomous process because the dialogue between machines.

Human beings have never been provided with so much data about every aspect of reality before and have never been given the ability to store, process and obtain information about them like nowadays (Delgado, 2017). We are living in a constant flow of information that feeds the economy, the legal framework, and the political activity of our current societies, defined by liquid values (Bauman, 2000), a hedonistic individualism (Harari, 2019; Vedder, 1999) and the fear that the 2020 covid pandemic has left us.

Our lives, security and development use the information that operates through the microdata and internet, so the data use protection is essential if it is associated to human personality in the private or public interactions. In this context, the law as a set of norms and institutions that orders life in society, protecting the values of freedom, equality, security and respect for human rights, must limit possible abuses in an area where machines can have enormous computing and information processing power without this translating into a decisive advantage for them or for those who operate them without democratically established controls. I suggest that even the best tools of artificial intelligence, cognitive computing, the Internet of things, the online work and even radical longevity should not be used to make certain decisions, which are less important than the protection of the dignity of the person (Leonhard, 2016).

Data, as attributes of material reality, can be perceived, stored and processed by living beings without conscience, through evolutionary phenomena. They are a part of their offerings, their ontology understood as the set of things that a living being can recognise and react to it. The behavioural environment constituted by all the relevant things for their wellness, in the sense that Dawkins (1994) asserts of "survival opportunities". It is a concept that Dennett (2017, pp. 80–81) names *umwelt,* following Jakob von Uexküll. In this sense, the legal concept of digital data and personal digital data provides the logical arguments to protect and guarantee the democratic human will as the power that operates in the decision-making processes and the fundamental human rights linked with privacy and freedom (Mazurier et al., 2019, 2020).

A digital data is a symbolic representation (numeric, alphabetic, algorithmic, etc.) of an attribute or qualitative or quantitative variable, in a digital support and it should be added that even though it is about the data format addressed to a machine (Cayón Peña., 2021).

It is in fact a datum addressed to a person, since all the described process is based on the human necessity to process information that constitutes his *umwelt.* The concept must include the massive storage of data and the industrial computation of data, especially the machine to machine processing data (M2M), because it could be considered personal data with a slight treatment relating each other (Paya et al., 2017).

The data to which the concept refers has a pretty varied origin, which entails that they also have a very varied format.

> According to IBM they can be classified according to their origin into the following types: - Internet and social media data: Facebook, LinkedIn, Instagram, Twitter, Blogs, Wikis and every kind of platform. - M2M data, that is the data that the machines automatically transmit to other machines. The volume is huge and it grows at a high speed. - Big Transaction Data that includes billing records, accounting, card operations or via Internet. - Biometrics that include from the retinal scan, fingerprint, heart rate data, facial recognition, genetics. - Human generators: call centre records, voice calls, electronic mails, photos and videos uploaded to the Internet. (Barranco, 2012).

The application of relationships between data can yield innocent personal information that we did not assume it could extract. To get an idea of the concept it is necessary to point out that the size of the big data is not defined, but according to the IBM, between a quintillion and 2.5 quintillion of data is generated every day coming from various sources such as clients, suppliers, financial operations online or obtained from mobile devices, social media analysis, GPS, etc. It is in this Big Data context where the exponential increase of the stored data volume and its processing affects the defence of the rights involved, as the mentioned privacy right increasingly enshrined in human rights conventions and other supranational legal instruments. (Paya & Delgado, 2017a).

In the Directive 95/46/EC about General Data Protection Regulation (GDPR) personal data processing should be designed to serve mankind and is defined as all information about an identified or identifiable natural person; that information by which a person's identity can be identified directly or indirectly, such as name, identification number, location data, an online identifier or one or more elements of physical, physiological, genetic, psychic, economic, cultural or social identity. Certainly, while big data presents many possibilities for good, it raises many moral and ethical concerns. The primary concern is an individual's right to privacy online. A critical point of view about it is needed, because the uncontrol commercial uses of big data has many interferences with individuals' rights (Friedman, 1997, 2002, 2013). Following Mantelero, 2018, p. 757.

> Moreover, human rights are largely safeguarded as individual rights, while Big Data and AI are often no longer primarily interested in the individual dimension and focus on groups and the collective level. For this reason, it is necessary to address the societal consequences of data-intensive applications, such as predictive policing or healthcare analytics (p. 757)

In the European legal framework, the protection of natural persons in relation to the processing of personal data is a fundamental right established in article 8 of the Charter of Fundamental Rights of the European Union and article 16 of the Treaty on the Functioning of the European Union (TFEU). It is not an absolute right; it must be considered in relation to its function in society and be balanced against other fundamental rights, in accordance with the principle of proportionality and as others fundamental and human's rights play according with the *Viena principles*. (Delgado & Giner, 2017).

In this human rights context, personal data protection, requires a stronger control by authorities and institutions, economic operators and public authorities should be enhanced. However, the industry of data seems to go in the opposite sense. Thus, the European Parliament has established in the art.4 of the Directive (EU) 2016/680 requiring all Member States to ensure that personal data are processed lawfully and fairly, collected for specified, explicit and legitimate purposes; adequate, relevant and not excessive; accurate and, where necessary, kept up to date; every reasonable step must be taken to ensure that personal data that is inaccurate, having regard to the purposes for which they are processed, are erased or rectified without delay; kept in a form which permits identification of data subjects for no longer than is necessary; processed in a manner that ensures appropriate security of the personal data, including protection against unauthorised or unlawful processing and against accidental loss, destruction or damage, using appropriate technical or organisational measures.

The strong cross-border dimension and a heterogeneous nature of personal data protection involves the public and private international law with subject as copy-rights, industrial property, and market and social tendences. In this scenario the North American model adds ideas to look for a real solution in guaranteeing the fundamental rights connected with digitalisation and its consequences, because the European model of public enforcement suffers from a "reality gap", or disconnection. In this sense Lehavi et al. (2016), about the north American model says.

> Does not give a fundamental right dimension to the protection of personal data, privacy and personal data is thus viewed primarily as goods that are at the disposal of private parties. To obtain a real protection both public and private entities, following their own models must work together to ensure better data protection, alongside the proper balancing of other interests, in the global digital era (p. 21).

2 Human and Fundamental Digital Rights, Getting the Future Right

Getting the future right—Artificial intelligence and fundamental rights is the title of 14 December 2020 reports of the Fundamental Rights Agency. It represents the European effort of creating a solution to the artificial intelligence and big data in human rights contexts. The fundamental right to data protection through the control of everyone's data, no matter which personal data, of its use and its assignment, are guaranteed to avoid the illicit trafficking in them or harmful to the dignity and rights of those affected. *As human and fundamental right the data protection right is configured as absolutely, unavailable, and universal. However, when we analyse the current measures and the consequences, an effective protection form must be developed in the involve rights protection and guarantee* (Delgado & Giner, 2017).

However, when we analyse the measures with which personal data is protected and in which way the owner is empowered with faculties, we understand that work must be done in the rights protection and guarantee. The legislation system that we have mentioned is, basically, to provide the data owner with the power to decide whether he wants to transfer their data or not and to be informed with the purpose of the use of this data. The protection system is completed through a variety of security measures that are regulated in the National Security Plan and some accessory rights from the owner: access, rectification, suppression, opposition and portability. The system establishes the consequences of the breach of the limitations imposed or the violation of the rights of the owner, that are basically concreted on sanctions to the person in charge and compensations to the data owner. The protection of a specific authority and the courts is also regulated but we are speaking about fundamental human rights and human dignity protection (Cayón Peña, 1998; Delgado & Giner, 2017).

In the use of internet, the person has no choice but to consent to the collection of the browsing data. The real utility of data is always yet to be discovered and it will have a future use that neither the entities nor us can advance when they request our consent for the granting of data. (Paya & Delgado, 2017b). Moreover, what is contemplated as closing measures are basically sanctions and compensation mechanisms. We cannot imagine how the damages from the unauthorised or improper use of data would be valued. (Ruiz-Ruano el at., 2019).

Certainly, it is a complex problem than could be resolved by the concept of algorithm, since it is defined as a sequence of logic steps for solving real data protection issues. It is possible to develop one *Troya* horse strategy for solving the juridical problem about the respect of data protection as fundamental rights and the use of internet and its capacities based in the human dignity protection and guarantee because is not only a fundamental right, as actual expression of human dignity in connection with the digital world (Halbertal, 2015) constitutes the real basis of fundamental rights, as stated in the Charter's official guiding explanations indicates.

In other words, human rights, big data and artificial intelligence are elements of a complex juridical algorithm, the use of the fundamental human rights as essential categories, its principles and recognition, based in the human person dignity, freedom, equality, and solidarity linked with the Design for Values, the Human Rights, Ethical and Social Impact Assessment (HRESIA model), of Mantelero (2018) and my *Dynamogenesis* of values and law (Mendez Rocasolano, 1996, 1999, 2013, 2014), could offer a possible answer to the demand for a more comprehensive assessment, including not only data protection, but also the effects of data use on other fundamental rights and freedoms in addition to social and political consequences.

According to Silveira and Rocasolano (2010), the dynamogenesis refers to the continuing process in which the values are immersed, and which can be summed up in the following steps: (1) knowledge-discovery of values by society; (2) eventual accession to social values and the immediate consequence; and (3) Implementation of the values through the law in its normative and institutional production.

Focusing on human rights as moral and social values, it is possible to design requirements that go beyond the present scope. In this sense the integration of the Design for Values with my *Dynamogenesis* of values and law made possible one gradual translation of abstract values into legal requirements and fundamental exigences of free, democratic and law states. Through the lens of human rights, annexing *HRESIA* model, to both models, the social and collective dimension of data uses results enabling, which is still not adequately addressed by the data protection regulations.

Van den Hoven et al. (2015) as I do with the environment right since the last century (Mendez Rocasolano, 1999, 2014). Using as I did three steps: in the "first, values are expanded into norms, which are properties or capabilities that the designed technologyhould exhibit to support desired values. For instance, the value privacy may in a certain context be interpreted as implying informed consent, confidentiality, and right to erasure rights" (p. 3).

In this sense the use of the *Dynamogenesis* of values and law could enforce the roots, principles and basis of the European and American models for develop authorities' resources, and a clear jurisdiction. The *Dynamogenesis* of values is like the Design for Values and explain the translation of moral and social values into a legal framework. See Aizenberg, and van den Hoven 2020 pp. 2–3 they explain how the ideas of "Friedman, 1997; Friedman & Hendry, 2019*;* Friedman et al., 2002*), Values in Design (*Flanagan et al., 2008; Nissenbaum, 2005*) and Participatory Design (*Schuler & Namioka, 1993 *over the past decades have opened up the design focus from technological artifacts and technical requirements to include a proactive consideration of the societal context in which the technology is embedded, and how societal needs and values can be translated into socio-technical design requirements*".

The *Dynamogenesis* of values and law is the continuous process in which values are immersed. Can be summarised in the following stages: (a) knowledge-discovery of them by society; (b) subsequent social adherence to them; and the immediate consequence: (c) concretisation of the values through law in its production.

In the first phase of knowledge-discovery of values by society, the valuational duty to be (Geistensollen von Werten) appears. Thus, when the Law regulates the conduct of the human community, reflecting through norms and institutions the axiological order. A good conduct is considered the adopted to the own social and cultural values, and the community rejects and penalises those who oppose it. These values in their pre-socio-legal and meta-socio-legal dimension are in an abstract space "the world of values" in this axiological world, the values are in suspense. Values do not exist because they are not felt, they have no worth. It is possible that the nobility of spirit of some human beings, makes them recognise and know, but until the social conscience does not recognise them, they are only valid for those who feel them. It is in a later phase, when the social adhesion to them arises, when the values of the latent axiological world come to life, it happens when the community feel and live them. When the social group demands them, values born into reality. Some of them will be valid, others will be rejected and ignored. We can say that the set of the valuable ones is the axiological feeling of the society (the community axiological order is created because it is valued, and for this reason, the human group feel and follow it).

In case of danger, the community will defend it passionately to safeguard their values. This social axiological feeling is represented in the constitutional rules. In the Spanish Constitutional Text, aims at just economic and social order within an advanced democratic society, in its article 1.1 where freedom, justice, equality and political pluralism are advocated as superior values of its legal system, and above them, presiding in article 10.1, the dignity of the person as the foundation of the political order and social peace. Already incorporated into the phase of concretisation of values through the Law in its normative and institutional production, the Constitutions manifest themselves, as a normative expression of social feeling. Law thus regulates social coexistence in a just manner by guaranteeing and protecting what society considers to be valuable. Through the values juridification, these values, which already are live, leap from the ideal (sentimental) to the real plane, because they can be demanded, guaranteed, and protected. Reality and effectiveness of the Social State are today present in the European discourse where the intention to protect the dignity of the person is alive, especially in those rights of intimacy which are affected by the protection of digitalized personal data, and which represent an attractive economic resource (Delgado et al., 2020a, 2020b). This idea is the soul that represents the essential content of the Social and Democratic States of law, since they represent not only a form of State, but also values that are the basis of European culture, such as Christianity and the conception of man with dignity. According to the dynamo genetic order, law captures the values "felt" as such by society and translates them into normativized axiological principles that are imposed through the rules of efficacy, and validity. What is socially valid is converted into a valuable duty to be.

3 Discussion

Big data and the artificial intelligence already play a role in the human deciding processes and for that important reason the democratic states need to make sure to fully uphold fundamental rights standards. If the human and fundamental rights most important goal is to protect the human dignity (Gerards, 2019; Raso et al., 2018), the big data and the IA gives an incredible tool to create differences, increase discrimination (Zuiderveen, 2018), control the human activities and control our will by one hand, but both develop an incredible world of a better way of living. (Harrison, 2011; Raso et al., 2018). For a peaceful coexistence with the human rights sets and exigences, the legal and political areas must work, as the algorithms do, being able to resolve complex aporias. Human rights law is based on the charters of fundamental rights, which provide the common basis for determining the use of data that is required in the context of the overall data processing policy. The Declaration of Human Rights, and the European Charter of fundamental human rights are guided by the values of human dignity, freedom, equality and solidarity in connection with the digitalisation, technological tools, applications and processes of use about personal data can play a helpful role in the century XXI development. Supranational courts and tribunals, which are entrusted with interpreting and enforcing such ideas, are playing a prominent role in giving substance to the human right to privacy (Docksey, 2020), designing a juridical architecture over the civil law and the old roman thesis of property rights, to protect the data rights, considering the specialty of the object on which it falls. The guarantee of the property right of our personal digital data is the power to decide over our lives using the concept of freedom and human dignity following the *Drittwirkung* theory we can decide over our rights (Böckenförde, 1993). The *mittelbare Drittwirkung*, understands that fundamental rights represent values and must, therefore, be guaranteed in private relationships assert that Drittwirkung already represents the primary function of fundamental rights (Suelmann, 1994). The real protection of privacy and personal data, as a right of present and future generations is perhaps one of the challenges of this new XXI century, where the artificial intelligence and big data could help to guarantee our fundamental rights with predictions and the analysis of the different possibilities and theories.

References

Aizenberg, E., & van den Hoven, J. (2020). Designing for human rights in AI. *Big Data & Society*. https://doi.org/10.1177/2053951720949566

Barranco Fragoso, R. (2012). ¿Qué es Big Data? IBM.com. https://www.ibm.com/developerworks/ssa/local/im/que-es-big-data/index.html

Bauman, Z. (2000). *Liquid modernity*. Blackwell.

Böckenförde, E. W. (1993). Sobre la situación de la dogmática de los derechos fundamentales tras 40 años de Ley Fundamental, cit., nota 28, pp. 114 y 115

Cayón Peña. J. (1998). El Derecho Internacional de los derechos humanos y el Derecho Internacional humanitario. *Boletín jurídico de la Universidad Europea de Madrid,* (1).

Cayón Peña. J. (2021). La seguridad de los datos en los ficheros universitarios frente a las amenazas actuales; análisis de riesgos y adopción de medidas de seguridad en relación con el cumplimiento del esquema nacional de seguridad. *Revista General de Derecho Administrativo*, (56).

Dawkins, R. (1994). *El gen egoísta*. Salvat.

Delgado Morán, J. J. (2017). Las relaciones internacionales del siglo XXI: transformar el mundo. Ed. Thomson Reuters.

Delgado Morán, J. J. & Giner Alegría., C.A. (2017). Equilibrio entre información y seguridad nacional o como el reforzamiento de la seguridad puede poner en duda el diseño del estado constitucional y democrático de derecho. en "El terrorismo como desafío a la seguridad Global". Ed. Thomson Reuters.

Delgado Morán, J.J., Jiménez Reina, J., Cremades Guisado, Á. (2020b). Analytical approach to emergent hybrid threats phenomena. Case Study: EU and Colombia. In: Ramírez J., Biziewski J. (eds) A Shift in the Security Paradigm.*Advanced Sciences and Technologies for Security Applications.*. Springer, Cham. https://doi.org/10.1007/978-3-030-43253-9_5

Delgado Morán, J. J., Jiménez Reina, J., & Jiménez Reina, R. (2020a). Seguridad cooperativa como medida de prevención y respuesta de la Unión Europea. Revista Científica General José María Córdova, 18(29), 61-85. https://doi.org/10.21830/19006586.520

Dennett, D. C. (2017). *De las bacterias a Bach, la evolución de la mente.* Ediciones de Pasado y Presente.

Docksey, C. (2020). The EU approach to the protection of rights in the digital environment: today and tomorrow-State obligations and responsibilities of private parties-GDPR rules on data protection, and what to expect from the upcoming ePrivacy regulation. pp. 47–78.

Flanagan, M., Howe, D. C., & Nissenbaum, H. (2008). Embodying values in technology: Theory and practice. In J. Van den Hoven & J. Weckert (Eds.), *Information Technology and Moral Philosophy* (pp. 322–353). Cambridge University Press.

Friedman, B., & Hendry, D. G. (2019). *Value sensitive design: Shaping technology with moral imagination.* MIT Press.

Friedman, B., Kahn, P. H., Borning, A. (2002) *Value sensitive design: Theory and methods.* UW CSE Technical Report 02-12-01. http://faculty.washington.edu/pkahn/articles/vsd-theory-methods-tr.pdf

Friedman, B., Kahn, P. H., & Borning, A. (2013). Value sensitive design and information systems. In N. Doorn, D. Schuurbiers & I. Van de Poel et al. (Eds.), *Early engagement and new technologies: Opening up the laboratory* (pp. 55–95). Springer.

Friedman, B. (Ed.). (1997). *Human values and the design of computer technology*. Cambridge University Press.

Gerards, J. (2019). The fundamental rights challenges of algorithms. *Quarterly of Human Rights, 37*(3), 205–209.

Halbertal, M. (2015) *Three concepts of human dignity*. https://youtu.be/FyEvREFZVvc

Harari, Y. N. (2019). *21 Lessons for the 21st Century*. Random House.

Harrison, J. (2011). Human rights measurement: Reflections on the current practice and future potential of human rights impact assessment. *Hum Rights Prac, 3*(2), 162–187.

Van den Hoven, J., Vermaas, P. E., & van de Poel, I. (2015). Design for Values: An Introduction. In J. van den Hoven, P. Vermaas & I. van de Poel (Eds.), *Handbook of ethics, values, and technological design*. Springer. https://doi.org/10.1007/978-94-007-6970-0_40

Lehavi, A., Larouche. P., Accetto. M, Nadya. N, & Zemer, L. (2016). *The human right to privacy and personal data protection: Local-to-Global governance in the digital era*. https://lawschool sgloballeague.com/wp-content/uploads/2017/01/Human-Rights-Group-Paper-2016.pdf

Leonhard, G. (2016). Technology vs. Humanity: The coming clash between man and machine. *Amazon Digital Services* LLC—KDP Print US

Mantelero, A. (2018). AI and big data: A blueprint for a human rights, social and ethical impact assessment. *Computer Law & Security Review, 34*(4), 754–772.

Mazurier, P. A., José Delgado Moran, J. J., & Paya-Santos, C. A. (2019). Gobernanza Constructivista de la Internet. *Teoría y Praxis*. Editorial Universidad Don Bosco, vol. 34, pp. 107–130

Mazurier, P.A., Delgado-Morán, J.J., Payá-Santos, C.A. (2020). The Meta-Tragedy of the Commons. Climate Change and the Securitization of the Arctic Region. In: Ramírez J., Biziewski J. (eds) Security and Defence in Europe. Advanced Sciences and Technologies for Security Applications. Springer, Cham. https://doi.org/10.1007/978-3-030-12293-5_5

Mendez Rocasolano, M. (1999). *El derecho a un entorno vital para el desarrollo de la persona*. Universidad Complutense de Madrid.

Mendez Rocasolano, M. (1996). Amparo, in: Prontuario de Derecho Constitucional, Pablo Lucas Verdú (org.) Comares.

Mendez Rocasolano, M. (2013). Perspectivas y perfiles de la cultura y el pacto de los derechos económicos, sociales y culturales a la luz de la fe y la justicia. *Prisma Jurídico, 12*(1), 51–93.

Mendez Rocasolano, M. (2014). La Teoría de la dinamogénesis de los valores y el Derecho a un Estado Social. Estudio preliminar del Traductor "Del Estado liberal al Estado Social" de Paulo Bonavides. https://www.researchgate.net/publication/281837214_La_Teoria_de_la_dinamogen esis_de_los_valores_y_el_Derecho_a_un_Estado_Social. Accessed 05 Sept 2021.

Nissenbaum, H. (2005). Values in technical design. In C. Mitcham (Ed.), Encyclopedia of Science, *Technology and Ethics*. Macmillan, pp. lxvi–lxx.

Payá Santos, C., & Delgado Morán, J. J. (2017a). Uncertainty of dimensional analysis of intelligence. *URVIO Revista Latinoamericana de Estudios de Seguridad, 21*, 225–239. https://doi.org/10.17141/urvio.21.2017.2962

Payá Santos, C., Cremades Guisado, Á., & Delgado Morán, J. J., (2017). El fenómeno de la ciberdelincuencia en España: La propuesta de la Universidad Nebrija en la capacitación de personal para la prevención y el tratamiento del ciberdelito. *Revista Policía Y Seguridad Pública, 7*(1), 237–270. https://doi.org/10.5377/rpsp.v7i1.4312

Payá-Santos, C.A., & Delgado-Morán J. J. (2017b) Use of cyberspace for terrorist purposes. In J. Ramírez & García-L. Segura (Eds.), *Cyberspace. Advanced sciences and technologies for security applications*. Springer, Cham. https://doi.org/10.1007/978-3-319-54975-0_12

Raso, F., Hilligoss, H., Krishnamurthy, V., Bavitz, C., & Kim, L. Y. (2018). Artificial intelligence & human rights: Opportunities & risks. *Berkman Klein Center Research Publication*, 6. https://doi.org/10.2139/ssrn.3259344

Ruiz-Ruano, A. –M., López-Puga, J., & Delgado-Morán, J. J. (2019). El componente social de la amenaza híbrida y su detección con modelos bayesianos. *Urvio. Revista Lationamericana de Estudios de Seguridad, 25*, 57–69. https://doi.org/10.17141/urvio.25.2019.3997

Schuler, D., & Namioka, A. (1993) *Participatory design: Principles and Practices*. L. Erlbaum Associates.

Silveira, V., & Rocasolano, M. (2010). Direitos Humanos: Conceitos. *Significados e Funções, 1*, 191.

Suelmann, H. G., (1994). Die Horizontalwirkung des Art. 3 II GG, Nomos Verlagsgesellschaft, p. 65.

Vedder, A. (1999). KDD: The challenge to individualism. *Ethics and Information Technology, 1*(4), 275–281.

Zuiderveen Borgesius, F. (2018). *Discrimination, artificial intelligence and algorithmic decision-making*. https://rm.coe.int/discrimination-artificial-intelligence-and-algorithmic-decision-making/1680925d73

María Méndez Rocasolano holds a PhD from the Complutense University of Madrid, a Fulbright scholarship from Harvard, and an honorary doctorate from the University of Toledo (Brazil). She currently teaches and researches as head of the Department of Constitutional Law at the UCAM Catholic University of Murcia. Spain.

The Challenge of Disinformation for National Security

Pablo Moral

Abstract This chapter provides an analysis of the implications of disinformation for national security. It examines the factors that make open societies particularly vulnerable, the actors that orchestrate disinformation campaigns and their strategic motivations. It also suggests possible ways forward to counter the challenge. It is argued that disinformation campaigns seek to undermine cohesion within democratic societies by sowing doubt, promoting confusion, and fostering division. The erosion of social trust aggravates political polarization, hinders governance, and provides fertile ground for extremisms. The sociopolitical circumstances in several Western societies, already polarized and immersed in the post-truth era, make them particularly predisposed to disinformation. Moreover, when used by states against foreign audiences, disinformation campaigns may aim at diminishing the international competitiveness of rival states, on the one hand, and produce on the other hand more amenable societies that ultimately influence their government's preferences towards ones more aligned with the disseminator's interests. Therefore, it is concluded that in the international arena disinformation is employed as a subtle, low-cost alternative to try to achieve gradual relative power gains at the expense of the rivals.

Keywords Disinformation · National security · Post-truth · Influence operations · Hybrid threats · International relations

1 Introduction

Over the last years, several Western democracies, including the US (Biden, 2021; Trump, 2017), Spain (DSN, 2017), France (Republique Française, 2015) and the UK (HM Government, 2015) have identified disinformation as a threat in their respective

P. Moral (✉)
PhD student, Pablo de Olavide University, Sevilla, Spain
e-mail: pmoral@lsi.uned.es

Researcher, National Distance Education University (UNED), Madrid, Spain

J. Cayón Peña (ed.), *Security and Defence: Ethical and Legal Challenges in the Face of Current Conflicts*, Advanced Sciences and Technologies for Security Applications,
https://doi.org/10.1007/978-3-030-95939-5_8

national security strategies. Accordingly, many countries as well as international organizations such as the European Union (European Commission, 2021) or NATO (NATO, 2020) have enacted laws or implemented measures against disinformation and external interference (Jeangène, 2021). The trend is clear: disinformation is a growing concern and states aim at improving their capabilities to protect from it. Despite being an ancient phenomenon (O' Shaugnessy, 2019), the new channels in which disinformation operates and the current political, social and communicational circumstances in which it develops have catapulted the challenge to a new dimension. Technological advancements made the spread of disinformation more accessible and far-reaching, whereas the sociopolitical context made individuals more prone to be targeted by disinformation campaigns.

The main goal of this chapter is to identify the implications of disinformation for national security. With that purpose, a qualitative analysis is carried out combining insights from the disciplines of Political Communication and International Relations (IR). Relying on academic and strategic literature and on multiple empirical precedents, the chapter examines who are the actors involved in disinformation campaigns, which strategic objectives they follow, and how their endeavors can be detrimental for national security.

This chapter unfolds as follows: in Sect. 2 a conceptualization of disinformation will be exposed, giving close attention, on the one hand, to the different conceptual frames in which it is usually incorporated, and on the other hand, to the factors that make it particularly harmful for open societies. Section 3 details the different types of broadcasters of disinformation, according to their structure and goals. After that, an examination of the strategic goals of disinformation campaigns is implemented in Sect. 4. Section 5 collects theoretical contributions from IR theory to distinguish how disinformation campaigns can affect global competition. Section 6 provides possible ways forward to tackle the challenge of countering disinformation. Lastly, a conclusion gathering up the main contributions and stressing some countermeasures against disinformation will close the chapter.

2 Disinformation in the Post-truth Era

Disinformation can be defined as false, inaccurate or misleading information that is intentionally disseminated, mainly to achieve political or economic objectives (Jack, 2017; Falis, 2015; Rodriguez Andrés, 2018). Therefore, disinformation is intentional: it is a deliberated endeavor that has a method and a purpose. This characteristic differentiates it from *misinformation*, which is false, inaccurate or misleading information shared without intending to deceive. Wardle (2020) suggests that misinformation and disinformation can include satire or parody, false connections and contexts, and misleading, imposter, manipulated and fabricated content.

Moreover, the lack of veracity distinguishes disinformation from *malinformation*, which is genuine information used against a target to cause harm—for example, leaks of private information—(Jack, 2017). It is plausible to interpret disinformation and malinformation as tools for deception and information manipulation. Following Carson (2009), deception involves successfully endeavoring to cause others to have false believes, which can be achieved, for example, by telling false or partially false statements, upholding information with the aim of misleading or telling a decontextualized truth. On its part, information manipulation occurs when the transmitter intentionally alters the quantity, quality, or the manner of the information with the intention to deceive, e.g., by telling the truth but not all the truth, or giving the message an intentional ambiguity in order to mislead (McCornack, 1992; McCornack et al., 2014).

Disinformation can also be included in some wide conceptualizations of propaganda (O'Saugnessy, 2019). For example, Jowell & O'Donnell's definition of propaganda corresponds to the purposive character of disinformation, although, by contrast, does not consider the veracity of the content. According to these authors, propaganda is "the deliberate, systematic attempt to shape perceptions, manipulate cognitions, and direct behavior to achieve a response that furthers the desired intent of the propagandist" (Jowell & O'Donnell, 2015, 7). Therefore, propaganda does not necessarily imply inaccurate or false content. However, in a more detailed approach, disinformation could be considered a form of *black* or *gray propaganda*, in which the information spread is false or inaccurate, according to Jowell and O'Donnell (2015). Besides, disinformation can be used as a device for psychological warfare, which involves the planned use of propaganda and other psychological operations to influence the opinions, emotions, attitudes, and behavior of opposition groups (Rand Corporation, n.d.).

In this chapter, disinformation campaigns are understood as coordinated activities disseminating disinformation with political or economic purposes. These can be carried out by foreign actors, implying hostile interference. Foreign disinformation campaigns are widely considered an element within hybrid threats, which are actions conducted by state or non-state actors, whose goal is to undermine or harm a target by combining overt and covert military and non-military means (Hybrid CoE, 2021). When synchronized strategically with other threats, disinformation campaigns can be employed as a tool in a *gray zone*, an ambiguous space between peace and war characterized by the use of multidimensional strategies—such as cyberattacks, economic coercion or military intimidation—that gradually pursue political objectives (Jordán, 2018; Mazarr, 2015).

The literature about the strategic use of information is not scarce. Terms such as *information warfare*, *information operations*, or *influence operations*, as well as disinformation, refer to disruptive information activities. Information warfare has a militaristic connotation, it mainly occurs in the cyber domain and although it has been usually associated with a context of conflict—for example, the Cold War—(Wanless & Pamment, 2019) this distinction is not always clear (see Nye, 2019; and Mazarr et al., 2019). Information operations are also predominantly used in the militaristic sphere to cover disruptive activities that affect the information and political system of an adversary (Wanless & Pamment, 2019). On its part, based on Pamment et al. (2018), Thomas et al. (2020), and Larson et al. (2009), influence operations can be defined as the organized use of information activities to influence perceptions, attitudes, behaviors, and decisions of foreign audiences in pursuit of a competitive advantage over an opponent. Disinformation is usually employed in influence operations, which can also incorporate truthful—at least to some degree—information.

Combined with other coercive measures or on its own, disinformation campaigns represent a challenge that is specially threatening for open societies. Firstly, these campaigns seek to exploit the democratic pillar of the free flow of information, and they are frequently developed in cyberspace, a domain that practically does not have borders and facilitates anonymity, making it difficult to identify the source. Secondly, disinformation campaigns usually occur in peacetime and do not overpass themselves the threshold that would justify a forceful response by the affected state. They cannot be considered "use of force" and the legal debate about if it violates the principle of non-intervention is opened (Sari, 2020). Thus, disseminators profit from moral and legal constraints of democratic states. And even though democratic states are determined to reinforce their legal and material capabilities, calibrating an appropriate response within democratic standards remains challenging.

Moreover, the sociopolitical context in certain Western democracies over the last years contribute to their vulnerability to disinformation campaigns. On the one hand, polarization and lack of trust in public institutions and traditional media outlets provide a fertile ground for manipulative attempts (Ingram, 2020). On the other hand, the progressive loss of influence of objective facts in shaping public opinion in favor of emotions and personal beliefs (Mcintyre, 2018) have made it easier for disinformation to appear plausible and effective. The post-truth era, or how Kavanagh and Rich (2018) put it, the "truth decay", refers to the blurring line between opinion and facts and the social disagreement about them, which can lead to the erosion of civil discourse, uncertainty, disengagement and political paralysis (Ibidem). Both references, Mcintyre (2018) and Kavanagh and Rich (2018), as well as this chapter, upheld a concept of truth that widely corresponds to the scientific truth, i.e., based on facts that can be empirically verified and tested. The abovementioned circumstances make today's democratic societies particularly vulnerable to disinformation campaigns, which entails implications for national security. Hostile actors, incurring little costs, can aim at causing public harm to predisposed audiences. The political stability and the international capacities of targeted states may thereby be affected.

3 The Perpetrators: Disinformation Actors

Following Melford and Fagan (2019), there are four types of actors behind disinformation campaigns, according to their degree of structure—centralized or decentralized—and their motivations—political or financial—. Within the centralized actors with political aims, two main groups can be distinguished: the state and the non-state actors. Jeangène et al. (2018) include within the non-state actors, on the one hand, the disinformation from terrorist groups, such as Daesh or Al-Qaeda. And on the other hand, they include the disinformation operations carried out by ethnic and religious groups, usually against minorities, such as the Saracen group, an online syndicate banned by Facebook for spreading fake news against non-muslim and Chinese population in Indonesia (Gleicher, 2019). Similarly, some political organizations can be added to this group of actors. Political leaders may incur in misleading messages to endorse their position, which might be considered a controversial manner within the normalcy of political competition. However, in some cases the organized and repetitive use of false and misleading statements and statistics may foster rejection and hate from part of the population towards political opponents or minorities. As an example, according to FBI data, hate crimes in the US almost tripled in the days following the 2016 elections (Williams, 2018), preceded by Donald Trump's inflammatory statements against ethnic minorities (Faber et al., 2017).

Concerning state actors, governmental authorities or agencies can be directly implicated in the design or spreading of disinformation. There is a long history of states resorting to disinformation for geopolitical purposes. For example, during the Cold War, Soviet Union's active measures against Western societies included the operation *infektion*, that spread the rumor that AIDS had been created by the US (Boghardt, 2009; Rid, 2020). The US also carried out disinformation campaigns in countries whose governments did not suit its geopolitical interests, like Guatemala, where Washington spread forged stories with the purpose of fostering instability in the Arbenz government.

A radio station established in Nicaragua by the CIA broadcasted fake dramatized examples of Communist tyranny and attempted to intimidate Arbenz's sympathizers (Grandin, 2007; Office of the Historian, 1975).

According to Bradshaw et al. (2021), in 2020 81 countries used social media for computational propaganda, defined as "the use of algorithms, automation, and human curation to purposefully manage and distribute misleading information over social media networks" (Woolley & Howard, 2019, 4), and to spread disinformation. These include major powers such as the US, China or Russia. These three countries, as well as fifty-four more, employed political bots. Furthermore, these authors found that in more than a dozen of states—involving again, among others, the three states just mentioned—there were "large numbers of staff and large budgetary expenditure on psychological operations or information warfare". In the same year, Twitter identified several influence operations carried out by inauthentic accounts allegedly linked to the governments of Iran, Russia, Venezuela, China, Thailand, Saudi Arabia, Turkey and Honduras (Twitter, 2021). Facebook, from 2017 to 2020, disclosed around a hundred "Coordinated Inauthentic Behavior" networks that came from fifty different countries (Facebook, 2021). However, some state disinformation campaigns are not covert and can be easily intuited. From the summer of 2019 onwards, China incremented its official presence on Twitter through the opening of dozens of diplomatic accounts all over the world. The intense and confrontational activity of many of these profiles made them be labelled as *Wolf Warriors diplomats* by Western analysts, in reference to a Chinese movie that praises China's expanding leadership role in international affairs and captures the country's growing nationalist sentiment. Among their misleading messages, some of these diplomats spread conspiracy theories about the origin of the virus—stating that it had been created

Table 1 Types of actors responsible for disinformation campaigns

	Centralized	Non-centralized
Political goals	State actors Non-state actors	Political trolls
Financial goals	Specialized private companies	Opportunistic rent seekers

Source adapted from Melford and Fagan (2019)

in a laboratory in the US, and at the same time stating that the virus was present in Europe before it got out of control in China—(Brand & Schafer, 2020; see Zhao, 2020) (Table 1).

State media outlets can be also considered state actors that may potentially broadcast disinformation. In these cases, the actions are not necessarily directed, ordered, or carried out by government representatives, but are expected to favor the state's strategic interests. The implication of the state is therefore not necessarily direct. RT—formerly Russia Today—is an example of state media that, with the aim of spreading their country's views further, broadcast highly biased content, potentially falling into misleading (Elswah & Howard, 2020) and manipulated content (see Maldito Bulo, 2021).

The non-centralized actors with political objectives are, in the words of Melford and Fagan (2019), the "grassroots trolls", that are defined as "individuals or groups that come together around a specific issue or cause". These groups lack structure and are often led by opportunism, rather than a specific agenda. They may intend to support state objectives, but motivations can be very varied and might involve hate speech or online harassment (Ibidem). As an example, the Covid-19 pandemic was utilized for spreading hoaxes demonizing political figures, famous personalities, refugees, and other ethnic minorities, such as the Chinese diaspora (Moral, 2020). According to Human Rights Watch (HRW), these stories led to an increase of hate incidents against for example Asian communities world-wide (HRW, 2020).

Within the centralized actors with economic purposes, Melford and Fagan include private influence operators, which are companies specialized in commercial marketing and public relations campaigns that are hired to disinform—as well as misinform and *malinform*—. According to Bradshaw et al. (2021) private firms increasingly provide manipulation campaigns, and between 2018 and 2020 there were more than 65 companies offering services to spread disinformation online. Among them, these authors spotlight the Israeli *Archimedes Group*, which ran campaigns in Africa, Latin America and South East Asia (DFRLab, 2019) and *Eliminalia*, a Spanish company that allegedly tried to interfere in local elections in some Latin American countries (Ángel, 2019).

Lastly, the fourth type of actors are non-centralized with financial objectives, or as Melford and Fagan call them, "rent seekers". They seek economic reward by attracting clicks, visitors, and ads through hyperpartisan and misleading content. Nonetheless, those well-structured media outlets that systematically spread disinformation for economic gains could well be considered centralized actors instead.

An example of non-centralized rent seekers is a group of Northern Macedonian teenagers that formed a lucrative small-scaled fake news industry about the 2016 US presidential election from a little town in their home country. Despite Cvetkovska et al. (2018) suggested a link between these teenagers and a pro-Republican local attorney linked to conservative politicians in the US, and even insinuated potential links with Russia, a later field research from Hughes and Waismel-Manor (2021) concludes that the "single goal" of local operators was to make a profit by fabricating false content and disseminating it among conservative Americans, who made their stories become viral.

4 Disinformation Strategic Goals

Disinformation campaigns seldomly seek to change previous beliefs in targeted societies (Torres-Soriano, 2019). By contrast, they usually seize the pre-existing prejudices and attitudes to manipulate the receptors. Indeed, disinformation exploits people's emotional and cognitive biases. Borrowing the definition of Haselton et al. (2015), cognition biases are "cases in which human cognition reliably produces representations that are systematically distorted compared to some aspect of objective reality". When this distortion is based on emotions (see Blanchette & Richards, 2010), it is referred as emotional bias. Among them, it is included the aversion to fear or the need for belonging to groups (Lin & Kerr, 2019), the confirmation bias, which leads us to believe information that corresponds to our preexisting views and attitudes (Lazer et al., 2018), or the motivated reasoning bias, through which we only tend to scrutinize thoroughly that information that does not correspond to our previous beliefs (Nemr & Gangware, 2019).

Internally, when a state utilizes disinformation towards its own population, it may aim at political control and the compliance of the population. Or, as O' Shaughnessy (2019) holds, at acquiescence. China's so-called *fifty cent army* is an example of disinformation to reinforce control and power. At least since 2004, thousands of internet commentators—who were rumored to be paid 50 cents (CNY)—were hired by the Chinese Communist Party (CCP) to praise the government, minimize criticism, and distract the attention away from sensitive issues (Han, 2015; King et al., 2017). State disinformation can also come from the ruling party to encourage people to follow a certain course of action. This is the case of certain members of the Republican Party and President Trump after 2020 US elections. Their unfounded accusations of electoral fraud—more than 50 lawsuits challenging the election were dismissed by the judges (Reuters Staff, 2021)—and conspirative stories against Democrats contributed to the radicalization of partisans, who, among other antidemocratic incidents, ended up attacking the Capitol (West, 2021).

The compliance of the population may also apply externally. Through disinformation campaigns foreign states do not aspire to suddenly become popular, but to rather seek the acceptance of international audiences (O' Shaughnessy, 2019). This could be considered a *soft* use of disinformation abroad. Nonetheless, disinformation can be weaponized against an opponent to weaken it and eventually obtain a competitive advantage. Indeed, one of the main aims of disinformation campaigns is to deepen the divisions of a targeted population (O' Shaughnessy, 2019; Jeangene et al., 2018; Nemr & Gangware, 2019). Disinformation may look for exploiting grievances, exacerbating tensions, and fostering polarization within a society, as well as damaging the state's relations with its allies. It may also seek to ignite the social debate by exacerbating emotions. A well-known example is the Russian interference in the campaign for US 2016 elections, according to the reports released by the Department of Justice (Mueller, 2019), the Office of the Director of National Intelligence (2017) and the House Permanent Select Committee on Intelligence (2018). A combination of hacks, leaks, disinformation, and trolls potentially linked to the Kremlin (Rid, 2020; Sanovich, 2019; @Policy, 2018) was accompanied by the militarization of microtargeting on social media (Colom-Piella, 2020), which allowed the perpetrators to deliver inflammatory content to specific groups and individuals based on their political stance, contributing to the radicalization of their positions.

Another goal of disinformation campaigns is to sow confusion in a society (Mazarr et al., 2019; O'Shaugnessy, 2019). One of the ways to do this is to promote plausible alternative narratives and false rumors, so the factual truth is relegated to be just one of many possible explanations of an issue (Torres-Soriano, 2020). This relativization of truth encounters fertile ground in the era of truth decay and reinforces it. The discredit of specific groups, public institutions, or personalities, such as elected politicians, scientists or media outlets also contribute to the confusion. For example, Chinese media and some Chinese diplomats on Twitter have spread misleading narratives about sensitive issues for the CCP, such as Hong Kong, Xinjiang or the Covid-19 pandemic, frequently attacking their geopolitical rivals and using networks of bots to amplify the messages (Wallis et al., 2020; Schliebs et al., 2021; Nimmo et al., 2020). A similar case is the attempt to undermine the credibility of rivals' achievements exhibited by the Sputnik V—Russia's main vaccine against Covid-19—official Twitter account, managed by the Kremlin. This profile, which tends to publish biased and misleading information, has even manipulated statistics about Western vaccines to spread mistrust and confusion (Buziashvili & Le Roux, 2021).

Hand in hand with confusion goes the aim of creating doubt in the targeted population (O' Shaugnessy, 2019; Ingram, 2020; Jeangène et al., 2018). There will always be a possibility of doubting the veracity of an issue (O' Shaugnessy, 2019). Firstly, because some matters cannot be totally proven correct (or incorrect), or at least not in a short period of time. The different not-proven scientific theories about the origin of the Covid-19 virus serve as a good example. The vacuum of information was soon filled by many forged narratives in different countries (Bandeira et al., 2021) that satisfied the demand for *cognitive closure* (Kruglanski & Webster, 1996), i.e., human's need to eliminate uncertainty and reach definite explanations. Secondly, even if there is scientific evidence about a fact, credibility can always be sabotaged by discrediting the source, criticizing the methodology, etc., hence a plausible rationale can always be found (O'Shaugnessy, 2019).

Division, confusion and doubt have the side effect of eroding trust within the affected population. Ingram (2020) proposes that what he calls "malign influence activities" weaken democracy by eroding the "trinity of trusts" in a society: social trust—defined as the belief in integrity and reliability of others—, trust in authorities—including experts—and trust in the democratic system. Eventually, the deterioration of trust provokes uncertainty and the perception of crisis in individuals. Following this author's line of thought, these circumstances might lead to the proposal of drastic solutions that are portrayed as appropriate and urgent, so non-democratic actions may be in this context perceived as legitimate and even necessary. In this way, part of the population might be seduced by non-democratic activities—even violence, in extreme cases—, or see a role model in foreign authoritarian leadership. Similarly, Webber et al. (2018) found empirical evidence on how extremist ideologies, by promising clear-cut strategies for restoration of certainty and personal significance, are appealing to individuals seeking cognitive closure.

Another eventual consequence of disinformation is the creation of suitable conditions for subsequent campaigns, since their goals are self-reinforcing. A polarized society with a high level of distrust is expected to be more vulnerable and therefore more open to further disinformation activities. According to Jeangène et al. (2018), the main factors of vulnerability to information manipulation are the presence of minorities, external and internal divisions, a vulnerable information environment and having contested institutions. Disinformation campaigns can seek to exploit these aspects to deepen the grievances and keep paving the way for further hostile action. Therefore, it can be inferred that those campaigns that are sustained over time are potentially more harmful.

5 The Strategic Logic of Disinformation in International Relations

International Relations theory provides appropriate lenses to decipher the utility of disinformation campaigns in the international arena. In particular the insights of Neoclassical Realism (NR) are found to be highly relevant. According to this approach, when carrying out foreign policy, states are not only influenced by the circumstances of the international system, but also by various internal variables. Ripsman et al. (2016) categorize these variables in four groups: leaders' perception, strategic culture, state-society relations, and national institutions. In democratic societies, these last two are particularly interconnected. State-society relations are defined as "the character of the interactions between the state's institutions and various economic and or societal groups" (Ibidem), and they are conditioned by the degree of cohesion there is within the society and between the society and the decision makers (Ibidem). National institutions determine how autonomously the state carries out the policies. In democratic systems, governments are subjected to institutional constraints, such as check and balances or parliamentary support, to implement foreign policy (Ibidem). According to Schweller (2006), the lack of consensus at the elite and the societal levels can lead to inappropriate responses to external threats

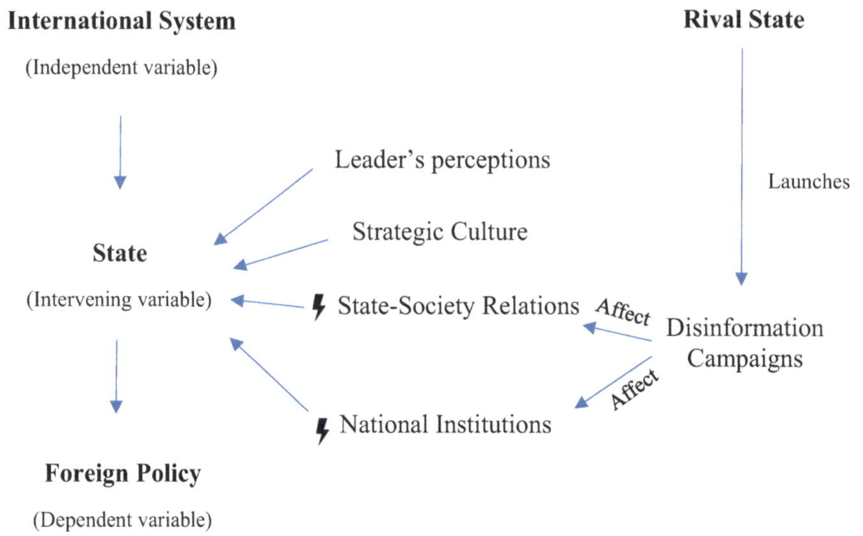

Fig. 1 The strategic logic of disinformation campaigns in IR, based on Neoclassical Realism. *Source* author. Based on Ripsman et al. (2016)

or opportunities, which may entail underbalancing and, therefore, an eventual loss of relative power. In highly polarized societies it is expected that leaders, instead of looking for the best options, take measures aimed at contenting as many people as possible, what Schweller calls "lowest common denominator policies" (Ibidem). In the same way, Ripsman (2009: 188) states that "when ambiguity or confusion reigns in the policy environment a domestic actor can most effectively emerge to shape policy" (Fig. 1).

Therefore, it is argued that NR provides plausible explanations for disinformation campaigns towards international audiences. Within the international arena, the information realm may be exploited as one of the domains of state competition and is meant to conduct further material gains in the medium and long term. When addressing foreign societies, the disseminator may aim at altering the state-society relations, which eventually would undermine national institutions (Fig. 1). The endeavors casting propagandistic content may seek to produce more amenable foreign societies that could eventually shape their state preferences towards ones more favorable to the perpetrator. Political, economic or military alliances might be facilitated, potentially converging towards what structural realist would call *latent power* gains: the socio-economic ingredients that go into building military power (Mearsheimer, 2001, 55). Moreover, by amplifying the societal grievances and discrediting their institutions and authorities, a state might expect that their geopolitical rivals face strong internal pressure. A weakened government, confronting strong opposition and dealing with polarized voices among the interest groups, might end up absorbed by domestic problems. Consequently, these circumstances can lead to a foreign policy that does not seize the opportunities efficiently and does not counter the threats appropriately, such as the rival's harmful strategies intended. Since foreign policy ends up affecting, in a longer run, the structure of the international system (Ripsman et al., 2016), disinformation campaigns may become a smooth, gradual, and low-cost alternative to seek relative power gains at the expense of the rival.

6 How to Fight Disinformation Campaigns in Democratic Societies

The first aid antidote against disinformation should be transparency (Torres-Soriano, 2020). Implementing transparent actions and processes would help to strengthen society's trust and reduce the margin for disinformation and conspiracy theories, which can arise from opacity, absence of data and lack of exemplarity of public representatives. Moreover, building confidence in the institutions is crucial for the steps against disinformation, which require the consent and collaboration and the civil society to succeed.

Furthermore, improving governmental capabilities must also become a priority. States should be equipped with permanent specific units integrated in national security structures. Examples of countries that have already set up such teams are the UK, Canada or Australia (Jeangène, 2021). As Milo and Klingova (2016) point out, these units should share and coordinate insights from different ministries and departments to efficiently tackle the multifaceted nature of the threat. Countries should be able to monitor potential threats, identify illegitimate informational activity, and expose the perpetrators publicly (when possible). The *name and shame*, even if it is not a definitive solution, might raise the public awareness of the danger and the need of being prepared against potential adversaries (Torres-Soriano, 2020). Governments should also have appropriate procedures to react within democratic parameters against external interference. Deterrence could be one of the ways to prevent an actor from launching disinformation campaigns (Pamment et al, 2018; Torres-Soriano, 2020). If the actor is clearly identified, political or economic sanctions can be implemented in order to raise the costs of an aggression. Moreover, states must have ways to encourage transparency and accountability of key actors, such as social media platforms and traditional media stakeholders, in order to advance towards a more honest and healthier media ecosystem.

However, in the long term there will be no good solution if people are not aware of their responsibility against disinformation. Empowering individuals is essential to its containment. Media and digital literacy, together with critical thinking are the shields through which society can resist deception and manipulation in social media (Pamment et al, 2018). Therefore, they should be incorporated into school curricula and, more urgently, extended beyond the younger generations. Indeed, Guess et al (2019) concluded that individuals over the age of 65 share seven times more fake news than the youngest cohorts, suggesting a significant gap among digital natives and older generations. It wouldn't be a matter of teaching people what is true or false, but of providing cognitive and technological devices that facilitate the identification of potentially biased, misleading, propagandistic or fallacious content.

In order to accomplish that, states need to ally themselves with multidisciplinary researchers, fact-checkers and specialized private organisms that in some cases have accumulated more expertise and technological capabilities than the governments (Polyakova and Fried, 2019). Examples of these institutions are the Atlantic Council Digital Forensic Research Lab, Alliance for Securing Democracy, EU Disinfo Lab, Graphika, Global Disinformation Index, The Oxford Internet Institute, Disinfo Cloud, Stanford Internet Observatory, Black Bird or First Draft. It is worth noting, that fact-checking, while useful in reducing misperceptions (Nieminen & Rapeli, 2019), is not a sustainable solution in the long term: it is a reactive, time-consuming activity which is not always effective and may not reach the appropriate audiences (Pamment et al., 2018; Pennycook & Rand, 2021). A related, promising way to prevent individuals from being manipulated is inoculation or pre-bunking, which consist in presenting information in advance so people can identify misleading content for themselves, such as accuracy prompts (Pennycook & Rand, 2021).

To keep improving these techniques, governments should support new pathways in science. The emerging multidisciplinary fields of Computational Social Sciences and Digital Humanities are key for the fight against disinformation, even if cross-area studies are usually not fostered in academia, where different scientific fields tend to work independently. These fields however are the drivers of artificial intelligence applications in the information sphere. For example, natural language processing is increasingly efficient in detecting inauthentic activity (San Martino et al, 2020) or manipulative language (Oshikawa et al., 2020). Promoting these multidisciplinary ways and private initiatives that apply them outside academia should be on every government's bucket list. These proactive bottom-up approaches, even if they do not always provide quick remedies, on the one hand tend to be less conflictive than reactive measures. For example, laws against disinformation, such the ones in Germany and France, found significant social contestation (Woitier, 2018), legal setbacks (Breeden, 2020) and criticism from NGOs (HRW, 2018). And the other hand, they address more directly the population's predisposition to further disinformation.

7 Conclusion

This chapter has analyzed the implications of disinformation campaigns for national security. With that purpose, it presented the factors of vulnerability in democratic societies and hostile actors that are willing to take advantage of this vulnerability. Moreover, the strategic goals of disinformation were discerned, as well as the logic of external interference in International Relations. Disinformation campaigns exploit and undermine fundamental aspects of democracy. It can be argued that they pervert the free flow of information and abuse the freedom of speech to erode mutual trust and coexistence. The instilled apprehension towards facts, towards authorities and even towards democracy aggravates polarization, contributes to political instability, and might eventually lead to the radicalization of individuals and civil unrest. These conditions might not only undermine the peace and prosperity internally, but also affect the state's international potential. In the long term, if the domestic political situation keeps being a drag on external action, the competitiveness of a state may be damaged in favor of its geopolitical rivals.

Tackling disinformation requires a comprehensive response that incorporates expertise from different areas, both at the governmental and the private institutional level. The multifaceted nature of disinformation campaigns involves technological, psychological, behavioral, sociopolitical, and economic implications that should not be addressed from silos of knowledge. Another remaining challenge that hinders the responses to disinformation is the difficulty in empirically measuring the individual effects of campaigns. It is not easy to discern to what extent social and political behavior can be attributed to disinformation. Likewise, it is arduous to accurately figure out the impact of the strategies taken against disinformation. Nonetheless, investing resources in multidisciplinary scientific research could be the way forward to improve capabilities and overcome current obstacles.

Acknowledgements This work was supported in part by the project Misinformation and Miscommunication in social media: bias (MISMIS-BIAS) under Grant PGC2018-096212-B-C32 from Spanish Ministry of Science.

References

@Policy (2018). *Update on Twitter's review of the 2016 US election*. Twitter Blog. Retrieved from https://blog.twitter.com/official/en_us/topics/company/2018/2016-election-update.html

Ángel, S. (2019). *¿Quiénes aparecen mencionados por empresa que estaría interfiriendo en elecciones locales?*. La FM. Retrieved from https://www.lafm.com.co/politica/quienes-aparecen-mencionados-por-empresa-que-estaria-interfiriendo-en-elecciones-locales

Bandeira, L. et al. (2021). *Weaponized: how rumors about covid-19 origins led to a narrative arms race*. DFR Lab. Retrieved from https://www.atlanticcouncil.org/wp-content/uploads/2021/02/Weaponized-How-rumors-about-COVID-19s-origins-led-to-a-narrative-arms-race.pdf

Biden, J. (2021). *Interim national security strategic guidance*. The White House. Retrieved from https://nssarchive.us/wp-content/uploads/2021/03/2021_Interim.pdf

Blanchette, I., & Richards, A. (2010). The influence of affect on higher level cognition: A review of research on interpretation, judgement, decision making and reasoning. *Cognition and Emotion, 24*(4), 561–595. https://doi.org/10.1080/02699930903132496

Boghardt, T. (2009). Operation Infektion. *Studies in Intelligence, 53*(4).

Bradshaw, S., Bailey, H., & Howard, P. (2021). *Global inventory of organized social media manipulation*. Oxford Internet Institute. Retrieved from https://demtech.oii.ox.ac.uk/wp-content/uploads/sites/127/2021/01/CyberTroop-Report-2020-v.2.pdf

Brand, J., & Schafer, B. (2020). *How China's 'wolf warrior' diplomats use and abuse Twitter*. Brookings. Retrieved from https://www.brookings.edu/techstream/how-chinas-wolf-warrior-diplomats-use-and-abuse-twitter/

Breeden, A. (2020). *French court strikes down most of online hate speech law*. The New York Times. Retrieved from https://www.nytimes.com/2020/06/18/world/europe/france-internet-hate-speech-regulation.html

Maldito Bulo (2021). No, esta foto de varias personas evacuadas de Afganistán llegando a Bélgica con armas en sus mochilas no es real: está manipulada. Retrieved from https://maldita.es/malditobulo/20210830/refugiados-afganos-armas-mochilas/

Buziashvili, E., & Le Roux, J. (2021). *How the Sputnik V Twitter account manipulates information to target Western COVID vaccines*. DFR Lab. Retrieved from https://medium.com/dfrlab/how-the-sputnik-v-twitter-account-manipulates-information-to-target-western-covid-vaccines-c1a13dac580d.

Carson, T. (2009). Lying, deception and related concepts. In C. Martin, (Ed.), *The philosophy of deception*. Oxford University Press.

Hybrid CoE (2021). *Hybrid Threats*. Retrieved from https://www.hybridcoe.fi/hybrid-threats/

Colom-Piella, G. (2020). Anatomía de la desinformación rusa. *Historia Y Comunicación Social, 25*(2), 473–480. https://doi.org/10.5209/hics.63373

European Commission. (2021). *Tackling online disinformation*. Retrieved from https://digital-strategy.ec.europa.eu/en/policies/online-disinformation

Rand Corporation. (n.d.). *Psychological Warfare*. https://www.rand.org/topics/psychological-warfare.html

Cvetkovska, S., Belford, A., Silverman, & C. Feder, L. (2018). *The secret players behind Macedonia's fake news sites*. Organized Crime and Corruption Reporting Project, July 18. Retrieved from www.occrp.org/en/spooksandspin/the-secret-players-behind-macedonias-fake-news-sites. GoogleScholar

Departamento de Seguridad Nacional (DSN). (2017). *Estrategia de Seguridad Nacional. Gobierno de España*. Retrieved from https://www.defensa.gob.es/Galerias/defensadocs/Estrategia_Seguriad_Nacional_2017.pdf

DFRLab (2019). *Inauthentic Israeli facebook assets target the world*. Retrieved from https://medium.com/dfrlab/inauthentic-israeli-facebook-assets-target-the-world-281ad7254264

Elswah, M., & Howard, P. (2020). Anything that causes chaos: The organizational behavior of Russia today (RT). *Journal of Communication, 70*(5), 623–645. https://doi.org/10.1093/joc/jqaa027

Faber, et al. (2017). Trump's electoral triumph: class, race, gender, and the hegemony of the polluter-industrial complex. *Capitalism Nature Socialism, 28*(1), 1–15. https://doi.org/10.1080/10455752.2017.1279867

Facebook. (2021). *Threat Report*. The State of Influence Operations 2017–2020. Retrieved from https://about.fb.com/wp-content/uploads/2021/05/IO-Threat-Report-May-20-2021.pdf

Fallis, D. (2015). What is disinformation. *Library Trends, 63*(3), 401–426.

Republique Française. (2015). Strategie National pour la sécurité numérique. Retrieved from https://www.ssi.gouv.fr/uploads/2015/10/strategie_nationale_securite_numerique_fr.pdf

Gleicher, N. (2019). *Taking down coordinated inauthentic behavior in Indonesia*. Facebook. Retrieved from https://about.fb.com/news/2019/01/taking-down-coordinated-inauthentic-behavior-in-indonesia/

Grandin, G. (2007). Off the beach: The United States, Latin America, and the cold war. In J. Agnew & R. Rosenzweig, (Eds.), *A Companion to Post-1945 America Malden*. MA: Blackwell Publishing Ltd.

Guess, A., Nagler, J., & Tucker, J. (2019). Less than you think: Prevalence and predictors of fake news dissemination on Facebook. *Science Advances, 5*(1), eaau4586.

Han, R. (2015). Manufacturing consent in cyberspace: China's "fifty- cent army." *Journal of Current Chinese Affairs, 44*(2), 105–134. https://doi.org/10.1177/186810261504400205

Haselton, M., Nettle, D., & Murray, D. (2015). The Evolution of Cognitive Bias. In D. Buss, (Ed.), *The handbook of evolutionary psychology*. Chapter 41. Wiley Online Library.

HM Government. (2015). *National security strategy and strategic defence and security review 2015*. Retrieved from https://assets.publishing.service.gov.uk/government/uploads/system/uploads/attachment_data/file/478933/52309_Cm_9161_NSS_SD_Review_web_only.pdf

House Permanent Select Committee on Intelligence. (2018). *Report on Russian Active Measures*. Retrieved from https://www.documentcloud.org/documents/4448600-House-Intel-Final-Report.html
HRW. (2018, February 14th). *Germany: Flawed social media law*. Retrieved from https://www.hrw.org/news/2018/02/14/germany-flawed-social-media-law
HRW. (2020, May 12th). *Covid-19 fueling Anti-Asian racism and xenophobia worldwide*. Retrieved from https://www.hrw.org/news/2020/05/12/covid-19-fueling-anti-asian-racism-and-xenophobia-worldwide
Hughes, H., & Waismel-Manor, I. (2021). The macedonian fake news industry and the 2016 US election. *PS: Political Science & Politics, 54*(1), 19–23. https://doi.org/10.1017/S1049096520000992
Ingram, H. (2020). The strategic logic of state and non-state malign 'influence activities.' *The RUSI Journal, 165*(1), 12–24.
Jack, C. (2017). *Lexicon of lies: Terms for problematic information*. Data & Society Publication. https://datasociety.net/pubs/oh/DataAndSociety_LexiconofLies.pdf
Jeangène, J. B., Escorcia, A., Guillaume, M., & Herrera, J. (2018). *Information manipulation: A challenge for our democracies*. CAPS-IRSEM
Jeangène, J. (2021). *Information defense: Policy measures taken against foreign information manipulation*. DFR Lab. Retrieved from https://www.atlanticcouncil.org/information-defense/
El conflicto internacional en la zona gris: una propuesta teórica desde la perspectiva del realismo ofensivo. *Revista Española de Ciencia Política, 48*, 129-151
Jowell, G., & O'Donnell, V. (2015). *Propaganda & persuasion* (6th edn). SAGE Publications.
Kavanagh, J., & Rich, M. (2018). *Truth Decay: A threat to policymaking and democracy*. Rand Corporation. Retrieved from https://www.rand.org/pubs/research_briefs/RB10002.html
King, G., Pan, J., & Roberts, M. (2017). How the Chinese government fabricates social media posts for strategic distraction, not engaged argument. *American Political Science Review, 111*(3), 484–501. https://doi.org/10.1017/S0003055417000144
Kruglanski, A., & Webster, D. (1996). *Motivated Closing of the Mind: "Seizing" and "Freezing"*. National Center for Biotechnology Information. Retrieved from https://www.ncbi.nlm.nih.gov/pubmed/8637961
Larson, E., et al. (2009). *Foundations of influence operations. A Framework for Enhancing Army Capabilities*. Rand Corporation.
Oshikawa et al. (2020). A survey on natural language processing for fake news detection. In *Proceedings of the 12th Conference on Language Resources and Evaluation (LREC 2020) European Language Resources Association*. Retrieved from https://aclanthology.org/2020.lrec-1.747/
Lazer, D., et al. (2018). The science of fake news. *Science, 359*, 1094–1096.
Lin, H., Kerr, J. (2019). On cyber-enabled information warfare and information operations. In forthcoming, *Oxford Handbook of Cybersecurity*, 2019. SSRN: https://ssrn.com/abstract=3015680
San Martino, et al. (2020). A survey on computational propaganda detection. arXiv preprint arXiv:2007.08024. https://arxiv.org/abs/2007.08024
Mazarr, M. et al. (2019). *Hostile social manipulation: present realities and emerging trends*. Rand Corporation.
Mazarr, M. (2015). *Mastering the gray zone: understanding a changing era of conflict*. U.S. Army War College Press.
McCornack, S., Morrison, K., Paik, J. E., Wisner, A. M., & Zhu, X. (2014). Information manipulation theory 2: A propositional theory of deceptive discourse production. *Journal of Language and Social Psychology, 33*, 348–377. https://doi.org/10.1177/0261927x14534656
McCornack, S.(1992). Information manipulation theory. *Communication Monographs, 59*, 1–16
Mcintyre, L. (2018). *Post-Truth*. The Mit Press
Mearsheimer, J. (2001). *The tragedy of great power politics*. Norton.

Melford, C., & Fagan. C. (2019). *Cutting the funding of disinformation: The ad-tech solution.* The Global Disinformation Index. Retrieved from https://disinformationindex.org/wp-content/uploads/2019/05/GDI_Report_Screen_AW2.pdf

Milo, D., & Klingová, K. (2016). *Countering information war: lessons learned from NATO and partner countries.* Globsec Policy Institute

Moral, P. (2020). La desinformación durante la pandemia de la covid-19 desde la perspectiva de los derechos humanos. En Hernández-Martínez et al. (eds.) *Derechos humanos ante los nuevos desafíos de la globalización.* Dykinson. pp. 1187–1201.

Mueller, R. (2019). *Report on the investigation into Russian interference in the 2016 presidential election.* US Department of Justice. Volume I of II. Washington, D.C. Retrieved from https://cop blaster.com/uploads/files/mueller-report_compressed.pdf

NATO. (2020). *NATO's approach to countering disinformation: a focus on COVID-19.* Retrieved from https://www.nato.int/cps/en/natohq/177273.htm

Nemr, C., & Gangware, W., (2019). *Weapons of mass distraction: Foreign state-sponsored disinformation in the digital age.* Park Advisors. Retrieved from https://static1.squarespace.com/static/5714561a01dbae161fa3cad1/t/5c9cb93724a694b834f23878/1553774904750/PA_WMD_Report_2019.pdf

Nieminen, S., & Rapeli, L. (2019). Fighting misperceptions and doubting journalists' objectivity: A review of fact-checking literature. *Political Studies Review, 17*, 296–309

Nimmo, et al. (2020). *Spamouflage goes to America: Pro-Chinese inauthentic network debuts english-language videos.* Graphika. Retrieved from https://public-assets.graphika.com/reports/graphika_report_spamouflage_goes_to_america.pdf

Nye, J. (2019). *Protecting democracy in an era of cyber information war.* Belfer Center for Science and International Affairs. Retrieved from https://www.belfercenter.org/sites/default/files/files/publication/ProtectingDemocracy.pdf

Office of the Director of National Intelligence, National Intelligence Council. (2017). *Assessing Russian Activities and Intentions in Recent US Elections.* Intelligence Community Assessment 2017–01D. Retrieved from https://www.dni.gov/files/documents/ICA_2017_01.pdf

Office of the Historian. (1975). 287. *Memorandum prepared in the central intelligence agency.* Retrieved from https://history.state.gov/historicaldocuments/frus1952-54Guat/d287

O'Shaughnessy, N. (2019). From Disinformation to Fake News: Forwards into the Past. In P. Baines, et al (Eds.) *The SAGE Handbook of Propaganda.* E-book edition. SAGE.

Pamment, J. et al. (2018, 1 July). *Countering Information Influence Activities: The State of the Art, version 1.4.* Lund University.

Pennycook, G., & Rand, D. (2021). The psychology of fake news. *Trends in Cognitive Sciences, 25*(5), 388–402. https://doi.org/10.1016/j.tics.2021.02.007

Polyakova, A., & Fried, D. (2019). *Democratic defense against disinformation 2.0. The brookings institution.* Retrieved from https://www.atlanticcouncil.org/in-depth-research-reports/report/dem ocratic-defense-against-disinformation-2-0/

Rid, T. (2020). *Active measures. The secret history of disinformation and political warfare.* New York: Farrar, Straus & Giroux.

Ripsman, N., Taliaferro, J., & Lobell, S. (2016). *Neoclassical Realist Theory of International Politics.* Oxford University Press.

Rispman, N. (2009). Neoclassical realism and domestic interest groups. In Lobell et al. (Eds.) *Neoclassical Realism, the state and foreign policy* (pp. 170–193). Cambridge University Press.

Rodríguez Andrés, R. (2018). "Fundamentos del concepto de desinformación como práctica manipuladora en la comunicación política y las relaciones internacionales". *Historia y Comunicación Social, 23*(1), 231–244. https://doi.org/10.5209/HICS.59843

Sanovich, S. (2019). Russia: the origins of digital misinformation. In C. Woolley, & P. Howard, (Eds.) *Computational Propaganda political parties, politicians, and political manipulation on social media.* Oxford University Press.

Sari, A. (2020). *Hybrid threats and the law: Concepts, trends and implications.* Hybrid CoE, Trend Report 3. Retrieved from https://www.hybridcoe.fi/wp-content/uploads/2020/07/Hybrid-CoE-Trend-Report-3.pdf

Schliebs, M., et al. (2021). *China's inauthentic UK twitter diplomacy: A coordinated network amplifying PRC diplomats.* Oxford Internet Institute. Dem. tech working paper 2021.2. Retrieved from https://demtech.oii.ox.ac.uk/research/posts/chinas-inauthentic-uk-twitter-diplomacy-a-coordinated-network-amplifying-prc-diplomats/#continue

Schweller, R. (2006) *Unanswered threats: Political constraints on the balance of power.* Princeton University Press.

Reuters Staff. (2021). *Fact check: Courts have dismissed multiple lawsuits of alleged electoral fraud presented by Trump campaign.* Reuters. Retrieved from https://www.reuters.com/article/uk-factcheck-courts-election-idUSKBN2AF1G1

Thomas, E., Thompson, N., & Wanless, A., (2020). The challenges of countering influence operations. *Partnership for countering influence operations, policy perspective series* 2. Canergie Endowment for international peace.

Torres-Soriano, M. (2019). A modo de introducción. La tormenta perfecta. In Torres-Soriano (coord.) *#Desinformación. Poder y manipulación en la era digital.* Comares.

Torres-Soriano, M. (2020). *Democracia vs. Desinformación. Propuestas para la protección de las sociedades abiertas.* Centro de Estudios Andaluces

Trump, D. (2017). *National Security Strategy (NSS).* The White House. Retrieved from http://nssarchive.us/wp-content/uploads/2020/04/2017.pdf.

Twitter. (2021). *Information Operations.* Retrieved from https://transparency.twitter.com/en/reports/information-operations.html

Wallis, J. et al. (2020). *Retweeting through the great firewall.* Policy Brief Report No. 33/2020. ASPI. Retrieved from https://www.aspi.org.au/report/retweeting-through-great-firewall

Wanless, A., & Pamment, J. (2019). How do you define a problem like influence? *Journal of Information Warfare, 18*(3), 1–14.

Wardle, C. (2020). *Understanding information disorder.* First Draft. Retrieved from https://firstdraftnews.org/long-form-article/understanding-information-disorder/

Webber, D., et al. (2018). The road to extremism: Field and experimental evidence that significance loss-induced need for closure fosters radicalization. *Journal of Personality and Social Psychology, 114*(2), 270–285. https://doi.org/10.1037/pspi0000111

West, D. (2021). *The role of misinformation in Trump's insurrection.* Brookings Institution. Retrieved from https://www.brookings.edu/blog/techtank/2021/01/11/the-role-of-misinformation-in-trumps-insurrection/

Williams, A. (2018). *Hate crimes rose the day after Trump was elected, FBI data show.* The Washington Post. Retrieved from https://www.washingtonpost.com/news/post-nation/wp/2018/03/23/hate-crimes-rose-the-day-after-trump-was-elected-fbi-data-show/

Woitier, C. (2018). *La loi contre les «fake news» voulue par Emmanuel Macron en fin votée.* Le Figaro. Retrieved from https://www.lefigaro.fr/medias/2018/11/20/20004-20181120ARTFIG00333-la-loi-anti-fake-news-voulue-par-emmanuel-macron-enfin-votee.php

Woolley, S., & Howard, P. (Eds.). (2019). *Computational propaganda: Political parties, political parties, politicians and political manipulation on social media* (p. 4). Oxford University Press.

Zhao, L [@zlj517]. (2020, March 13th). This article is very much important to each and every one of us. Please read and retweet it. COVID-19: Further Evidence that the Virus Originated in the US. https://globalresearch.ca/covid-19-further-evidence-virus-originated-us/5706078… [Tweet]. Twitter. Retrieved from https://twitter.com/zlj517/status/1238269193427906560

Ethics and Law Facing Other Current Conflicts

Ethical Behaviour and Social Responsability from the Perspective of Modern Psychology

Marzanna Farnicka

Abstract The article discusses problems with choosing rational and socially responsible behaviour when facing danger, fear or threats. This issue is extremely important nowadays, particularly when we witness conflicts and recognize threats (physical, mental, economic, health threats or natural disasters) in the immediate vicinity and cyberspace including mass and social media. Currently any large-scale threat can be created in a matter of hours and it is difficult to distinguish real threats from media-created or virtual ones. All these threats may reduce individual well-being. However, regardless of the source of the conflict or the situation identified as a threat, the reactions and behavioural tendencies of small groups and individuals remain constant. A number of studies have shown that these include role-taking (Standford Experiment), submission to authority (Millgram Experiment), coping processes include succumbing to prejudices, stereotypes, emotions, dichotomising the world, narrowing cognitive schemas or the inability to resist temptation. This article discusses question relevant to security sciences and security studies[i] from positive psychology's perspective: whether it is possible to control or manage individuals and small groups through fear, reducing people's reactions to basic mechanisms of fight or flight response, mental and rational freezing, as well as freezing linked to actions taken or responsibility. The confrontation of contemporary knowledge about human reactions in crisis situations makes us reflect on whether we are really doomed to them? How to deal with them regardless of biological limitations? How to deal with the illusion of twentieth century idealism? What procedures can be used to ensure that engaging in responsible behaviour does not infringe on the freedom of individuals and groups? These questions point to new areas of research and reflection.

Keywords Idealism · Axiology · Human rights · Social experiments · Social impact · Neuroscience · Trauma · Values · Well-being

---

M. Farnicka (✉)
Institute of Psychology, Zielona Góra University, Zielona Gora, Poland
e-mail: m.farnicka@wpsnz.uz.zgora.pl

J. Cayón Peña (ed.), *Security and Defence: Ethical and Legal Challenges in the Face of Current Conflicts*, Advanced Sciences and Technologies for Security Applications,
https://doi.org/10.1007/978-3-030-95939-5_9

1 Past Ideas—Social Psychology Experiments

In retrospect, the twentieth century can be called the era of the loss and regaining of optimism and faith in humanism. The beginning of the century was the triumph of the Enlightenment, which was based on the values of reason and progress. Innovations such as the airplane (constructed by the Wright brothers, 1903), the vacuum cleaner, the automobile, the camera and nuclear power were associated with this period (Buchanan & Erdelez, 2019). In the first half of the twentieth century, Einstein developed his theory of relativity. Lenin attempted to spread his idea of communism (February 1905 and October 1918 revolutions: Lukas, 2013). This period was marked by uncertainty caused by progressive discoveries. Radical new ideas and ways of thinking were proposed that dramatically challenged the traditional order of things. Revolutions, First and Second World War erased the optimism about human nature in several countries across the world.

After the Second World War, people started dreaming of a new deal, global peace and human rights. Immediately after Second World War, massive inequalities became apparent ranging from racial and gender discrimination to political persecution. These movements and their effects made the 1960s a progressive decade. To deal with the past, psychological experiments were conducted in order to explain human behaviour beyond morality. A number of social experiments attempted to identify the sources of unethical and unlawful behaviour. The first influential experiment that revolutionised the view of human behaviour was the Sherif experiment (1936) followed by experiment conducted by Asch (1956). The following studies were carried out by teams leaded by Milgram (1963) and Zimbardo (1970).

1.1 Asch Conformity Experiment

Based on Sherif (1936, 1956) experiment in which the mechanism of transmission of social norms from generation to generation was inferred. Asch (1956) disagreed with Sherif's interpretations of his study participants' behaviour, claiming that participants were only inclined to yield to the group to a large extent when they feel insecure and had no formed opinion on a given topic. Asch expected that under unambiguous conditions, participants would not be submissive to the group.[ii]

The researcher was surprised to find such high percentages of people conforming to the group's answer in the absence of clear group pressure. Asch interpreted this result as the symptom of submissiveness to the majority—conformity—motivated by the fear of rejection by the group and the desire to be accepted by group members. This is the main motive of so-called "normative conformity". Such conformity is based on the tendency to conform to the unwritten norms of the group for fear of being excluded. The second type of conformity Ash pointed to is cognitive conformity, which concerns a not fully conscious change in perception in a situation of reality recognition. Cognitive conformity may arise as a result of normative conformity.

What is important in this case is that an individual does not refer to norms but to his or her own perceptions (Asch, 1956).

1.2 Milgram Experiment

Millgram (1963) was another researcher who investigated the reasons for acting contrary to accepted norms and yielding obedience to even cruel commands. Miligram tested the hypothesis of whether obedience that takes away one's common sense has social or individual causes by conducting an experiment to investigate people's tendency to obey authority figures. In the first series of experiments, 40 individuals between the age of 20 and 50 were studied: fifteen of them were workers, sixteen were sales representatives or entrepreneurs and nine other professions. Each of the participants was tested individually. When the participant entered the lab, he or she saw another 'tested individual' already present in the room. A faked role-drawing took place in which the subject of experiment was assigned the role of "teacher" and the experimenter's assistant the role of „learner". The teacher was told that his or her task would be to read a list of word pairs (e.g. "nice-day", "blue-box") to the learner and then check how many of these pairs the learner had memorised. The checking phase was to have the teacher read the first word from the pair, then four suggestion words, from which the learner must select the correct word (by pressing one of the four buttons).

If the learner answered correctly, the teacher would read the next word and if the student answered incorrectly, the teacher would administer the electric shock to the learner as a form of punishment.

The teacher was initially given one trial electric shock of 45 V, which was felt to be quite painful.

The rule was that with subsequent mistakes the strength of the shock was increased. The initial dose was 15 V. Each subsequent shock was supposedly 15 V stronger. The final dose was supposed to be 450 V—there were 30 switches in front of the teacher.[iii]

As a result, 26 out of 40 participants completed the experiment applying 450 V shock to the learner.

The results did not vary across participants' age, gender or nationality.

Due to the fact that such a high percentage of total obedient participants was unexpected, the experiment was repeated in several variations on several nationalities including Duch, German, Spanish, Italian, Australian and Jordanian (Meuus & Raaijmakers, 1986) over the years. It was found that the propensity to obey was similar across countries and continents. The results of subsequent studies indicated that the problem of obedience depends largely on situational rather than personality factors.

1.3 Stanford Prison Experiment

The last significant research indicating social conditioning of human behavior was the "prison experiment" conducted by Ph. Zimbardo with his team. In this study, students randomly played the role of guards and prisoners. This experiment confirmed that mentally healthy people in a specific conditions can successfully play the roles of perpetrators and victims. Zimbardo (2007) argues that the reason for such behaviour is not a mental disorder but the influence of the environment on the individual, particularly when the individual does not have the opportunity to abandon the role schema or when the social role does not allow for a margin of freedom of behaviour (although the experiment was initially planned for a fortnight, was interrupted after only six days due to the brutal behaviour of those who took the roles of guards). Importantly, it was shown that not only the role but also specific attributes fostered a group process in which violence escalated. These attributes were unusual clothes of the prisoners (they were dressed in long white shirts with an identification number on both sides, a padlocked chain was attached to the right foot of each prisoner, their cap was a woman's stocking), prisoners' de-individuation (all prisoners looked identically apart from having different identification number) and specific communication code implemented (the prisoners had to call the guards "Mr Penitentiary Officer" and to each other only by their identification number) (Haney & Zimbardo, 1976; Zimbardo, et al. 1999).

1.4 Summary—Hope for a Better World

The state of knowledge after „the golden times of big social experiments" may be summarized that they identified factors shaping submissive behaviour towards authorities and roles, as well as factors shaping critical and independent attitudes. At that time, the importance of particular factors of social situations and processes was emphasized. Factors that intensified the submissive attitudes towards unethical and cruel rules of the subjects were:

- the proximity of the authority and the degree to which it is accepted as the "real authority" (charisma),
- institutionalisation of the authority; if the experimenter was perceived as an employee of a university, the subjects' obedience increased (the similar phenomenon could be observed when another institution stands behind the person, e.g. bureau, army or hospital),
- lack of direct contact or a long distance between the subject of experiment and a victim,
- authoritarianism of subject of experiment (characterised by the worship of authorities)—the authoritarian subjects were slightly more obedient (they were more likely to use maximum electric shocks). However, this personality influence was not significant.

In contrast, obedience was reduced by factors such as:

- the inability to avoid responsibility for the victim's suffering,
- contradictory orders from various authority figures,
- the presence of other individuals who oppose the commands of authority (allies),
- either physical or emotional proximity to the victim.

Even almost all experiments had many ethical problems because of they consequences to the participants (Milgram, 1974, Zimbardo et al., 1999) such a state of knowledge gave great hope for changes in the society. The role of education and creation of the right conditions for human development was highly emphasized (Bandura et al., 1975; Bond & Bushman, 2017). For example the Peyrefitte with a special Committee prepared the report (1977). It aimed to diagnose the conditions of the emerging sense of insecurity and to indicate the most important and necessary actions to be taken to counteract pathological phenomena such as violence, crime, and delinquency. The process of inducing insecurity was analyzed from an individual perspective as an act of violating the individual's well-being. The Report set out the steps that a modern country should take to protect society from individual, institutional, economic or social violence.

The main controversy that appeared during the preparation of the Report and its 105 recommendations aimed at stopping the development of violence and crime, was the issue of the death penalty. The Committee was explicitly for its abolition as one of many conditions for improving the quality of the life of societies. This issue became central because it affected the values that guided the authors of the Report. They referred to the idea of humanism as the main principle behind the organization of modern societies.

2 Times of Doubt

In last decades of 20th Century, scientific and technical development, and the explosion of optimism and new ideas turned into social problems and cultural revolutions. People witnessed the technological changes (internet invention, cell phones, new medical treatments and technology) and believed in human power (eg. in explosion of humanistic psychology, cognitive—behavioural psychology), individualism and perpetration. Cultural relativism extended, by which Western societies no longer saw themselves as having a superior culture. Mass culture, a product of mass media and the progressive availability of leisure time led to a new artistic sensitivity. In the end of twentieth century, we have received the idea of positive psychology stating that all human beings and creatures have positive aim of their existence (Maslow, 1966, Seligman, 2011). On the other hand, a new in-depth era of research on the human brain began. The researchers not only searched for genes responsible for aggression but also attempted to discover specific factors which trigger particular behaviours. Through the development of behavioural engineering, we entered the twenty-first century with the hope of developing behavioural genetics.

At the end of the twentieth century there explosive moment of modern times—liberation (modernism) occurred including: political liberation, sexual liberation, liberation of productive forces, liberation of destructive forces, liberation of women, children, unconscious impulses, liberation of art, education and science (Baudrillard, 1992). Along with technical, economic, and social changes, we could observe the axiological and cultural changes (Appadurai, 2013). Some ideas resist more than others by enacting laws for the preservation and protection of traditional cultural patterns while putting up barriers to alien ideas (Edmunds & Turner, 2005). As an example the document called the Seville Statement on Violence (SSV) was presented (SSV, 1990). The SSV based on gathered results of study about sources of aggression prepared by international scientist. SSV was adopted in the international meeting of scientists, convened by the Spanish National Commission for UNESCO in Seville, Spain 16th May 1986. The SSV as a document was subsequently adopted by UNESCO at the twenty-fifth session of the General Conference on 16th November 1989. The statement known as a 'Statement on Violence', was designed to refute "the notion that organized human violence is biologically determined". The belief about the relationship between warfare and human biological conditions may have practical implications. It has been proved that if an individual believes that war is biologically determined, he or she is less likely to engage in activities to promote peace. In 1997, Pinker has used the Seville Statement as an example of the idea of biological determinism, the incorrect idea that genes are solely responsible for *any* of individual' behaviors. The evolutionary point of view (Jones, 2008) underlines that violence is not done for its own sake but is a by-product of goals such as higher status or reproductive success.[iv] Because of social needs in XXI century the researches associated with International Society for Research on Aggression (ISRA) once more summarised results of study on risk factors for youth violence (Bushman et al., 2018). What is important in that raport the causes of violent behavior are treated very complex. Authors separated known risk factors for youth violence into two categories: (1) personal risk factors associated with the individual, and (2) environmental factors associated with the situation or broader social environment.

2.1 XX Century—Process of Changing Values of Social Responsibility

To summarise, in the twentieth century, we have gained concrete scientific knowledge about the conditions of 'good, pro-social behaviour' and the conditions of negative, selfish, aggressive and violent behaviour. Researchers defined and examined many processes and phenomenas such as informational conformity (Sherif), normative conformity (Ash), the phenomenon of group pressure (Ash), obedience to authority (Millgram) and the group process that causes identification with a role (Zimbardao) and intra and inter group processes (Jones). There were also identified several factors causing the escalation of unwanted behavior including: insecurity and need for acceptance (Ash), proximity to authority and a degree of acceptance as "real authority" (charisma), institutionalisation of authority (Millgram), lack of direct contact or a distance from the victim and depersonalisation of others (setting up an us-them conflict) (Millgram; Zimbardo) or process of modeling (Bandura). In that times, based on the achievements, many education programs to increase level of social life were established (Brännström et al., 2016; King, 1988).

The level of ethical behaviour in accordance with accepted norms is increased by the awareness of ability to influence the situation, an inability to avoid responsibility for the suffering of the victim and conflicting orders from different authority figures and proximity to the victim (either physical or emotional).

3 What Happened at the Beginning of 21st Century?

At the beginning of twenty-first century, the knowledge about human existence and psychological processes was extended. The predominance of genes, brainpower, and a technological context make us a global village with the illusion that we are all the same. This is largely true, however the core of the progress of psychology as a science that predicts and explains human behaviour is in the undisclosed and not well-explained differences. Biological sciences focus on examining the existing knowledge and demonstrating how existing achievements are based on individual and species-specific biological factors. We all were the witnesses of technological revolution (new media, new cyber environment, the new possibilities of transport, communication and treatment) (Abramova & Krasheninnikov, 2018).

3.1 Between Biological Determination and Individualistic Freedom

What is essential for this ages is the fact that mechanisms of biological responses and their determinants due to technology possibilities have been once more defined, explained and underlined. Neuroscience studies have found that as a result of experience, real physiological changes occur in our nervous system particularly in memory, senses, level of readiness to react, calibration of the brain's warning system, activity of stress hormones (Davidson & Begley, 2012; Seo, et al., 2021) There were also another changes found in the information-selecting system, in the system responsible for self-regulation and the ability to pay attention (Kandel, 2006; Nigg, 2017). The current knowledge explains why individuals experiencing various types of trauma (severe stress, negligence, abandonment, violence or life-threatening), become overly sensitive to threats at the expense of spontaneous involvement in daily activities. The appropriate treatment is essential in order to overcome behaviour patterns of attack (fight response), withdrawal (flight) or indifference (numbness) (Damasio,

2000; Eisenberg et al., 2010). These all works underlined that individuals experiencing trauma are unable to perceive reality adequately, learn by one's mistakes, have lower levels of empathy and are mainly focused on survival or preserving the past (Blaustein & Kinninburgh, 2012; Zeanah & Sonuga-Barke, 2016).

The contemporary neuroscience has provided new insights into the development of morality and ethics and their foundations. The nature of relationships with other people from the first days determines the development of several functions that are recorded in the nervous system at physiological, morphological and functional levels (developmental trauma) (Parera et al., 2020). The contemporary knowledge explains the meaning of adverse events (war traumas, accident traumas) or significant family events not only on an individual level but also on a generational level (Guerrero et al., 2020).

The knowledge of the basic processes and mechanisms underlying behaviour has opened up new possibilities for understanding, mitigating and reversing the effects of damage and problems in response to reality. Researches on the role of emotions and functioning in social situations (LeDoux, 2012; Zajonc, 2000) have enabled the examination of the development of hostile, aggressive or destructive behaviours. A number of studies, for example Chenung & Ju (2016) have highlighted the importance of feeling anxiety in organising information about others. Alice LoCicero's research presents the developmental perspective of perpetrators (Lo Cicero, 2014, 2016). Recent studies show that social and emotional processes are present in a virtual space too (Chenung et al., 2019, 2021). It is important that for the brain is difficult to distinguish real threats from media-created or virtual ones (Bond, & Bushman, 2017; Dillon, & Bushman, 2017). Advanced technology allows to search and present the sources of unethical and immoral behaviour a result of real physiological changes in the brain of an individual (Ba & Bhopal, 2017; Farnicka, 2017; Gabor, 2011).

3.2 Return to Humanism and Individualistic Development - Positive Psychology Framework

It should be emphasised that in the twenty-first century there is a growing trend towards a holistic approach to the individual. The current state of physical science allows us to understand the nature of brain waves and to see emotions in the brain in the form of an EEG, MRI and in the form of colours and electromagnetic spectrum. On the basis of these technological possibilities, the positive psychology is developing dynamically. Similarly to the humanistic psychology (Maslow, 1966) in the twentieth century, the positive psychology avoids being a limited paradigm and aims to be a metatheory rather than a limited paradigm. Positive psychology is understood as a theoretical and empirical trend in which the attention of researchers is focused on the abilities and skills of individuals that favour taking up challenges, achieving a high quality of life and coping with difficult situations. Positive psychology aims to perceive an individual as a creative and productive being, enjoying life and coping with adversities (Trzebińska, 2008).

The three research pillars of positive psychology are intended to fulfil the above tasks (Seligman, 2011). The first pillar refers to the research on an individual's positive, subjective experience of life in the past, present and future. The positive experience of the past can be described through contentment, satisfaction, and well-being, while the experience of the present can be described through concepts such as happiness, flow, flourishing or sensory pleasure. Finally, a positive experience of the future can be reflected in terms of optimism and hope. The second pillar of refers to the study of the so-called good personality, which is a system of timeless and transcultural virtues that include wisdom, humanity, courage, restraint, justice and transcendence. This pillar also includes research on the strong traits such as curiosity, consideration, forgiveness that condition the emergence of virtues. The third pillar involves the analysis of positive society. In this area, researchers identify institutions such as democratic system in the country or a properly functioning family, that enable individuals to develop positive mindset. As highlighted by Seligman (2003), social institutions and the processes occurring within them support the abilities of individuals (e.g. involvement, responsibility), and thus create conditions for the emergence of positive emotions, which are an essential factor in coping with difficult situations.

The concept of positive psychology also identifies possible dimensions of change and directions of development. Seligman (2003) presented a model which consists of five elements that make life good. These are (1) P—Positive emotions (the ability to focus on positive emotions, e.g. joy, recognition, comfort, hope, curiosity); (2) E—Engagement (state resulting from having competencies to achieve goals and motivation to fully devote oneself to action); 3) R—Relationships (being among people, positive and supportive social relations—cooperation, friendship, love); (4) M—Meaning (sense of meaning in life and one's participation in various spheres of activity); (5) A—Accomplishment /Achievement. (achievement of goals and plans). The model called PERMA and it postulated that each component contributes to well-being, is pursued for its own sake, and is defined and measured independently.

The another model of Keyes (Keyes & Waterman, 2003) highlights the necessity of perceiving individual's behaviour in a specific context (social wellbeing). This includes social acceptance (trusting society and conviction of its sympathy towards an individual), social coherence (perceiving the world as understandable, predictable and orderly), social actualization (belief in the positive evolution of society), social integration (the sense of the individual that one is part of society and is supported by it) and social contribution (the belief that an individual is integral part of the society). These concepts and the possibility to measure them enable interdisciplinary cooperation in forming security within security sciences or creating a culture of security (Piwowarski, 2020).

4 What Can We Expect in the Future?

It may appear that the great discoveries of neuroscience have revolutionised our treatment of human beings. The industry is using all possibilities of application of this knowledge. Neuroscience is used in medicine (e.g. in addiction therapy—EMRD method, treatment of depression), education (neurodidactics, cyber-education), marketing, economics, human resource management and leisure (games developing) (Alpert et al., 2021). It should be stressed that high technologies changes not only in the field for which they were developed. When predicting the development of society, it is necessary to take into account a wider range of their impact on the entire socio-cultural system as a whole and each individual in particular.

In the third decade of the twenty-first century, we are developing methods that put into practice the natural flexibility of human brain (discovered in) in order to help individuals with different maladaptive experiences to anchor themselves in the present and to help perceiving the reality not only through a prism of threat and rejection (Gruber et al., 2014).

According to Van der Kolk (2014), treatment and optimisation of behaviour occurs through three paths. The first path tackles the problem of relationships by establishing relationships or rebuilding bonds with others and teaching ways of understanding and taking control in situations that resemble traumatic experiences and trigger behavioural patterns of fighting for life (this applies to physical, social or psychological aspects). The second path is through medical solutions. Nowadays it could be to observe an incredible development in the pharmaceutical sector which gives us hope that drugs are able to block abnormal alert reactions.

This can change the way information is organized in the brain, or allow the experience of other emotions (anxiety, fear, anger) which affect the perception of reality and the behaviour that is triggered. The third path is through the experience of feelings from extremely deeply felt powerlessness of rage or helplessness caused by trauma and the correction of these states.

4.1 What Are We Left with?

As a human, we have received highly idealistic leads for the future: reducing inequalities, providing more available resources, and reducing blood feuds due to better functioning justice systems, which may have contributed to declining intra-group violence.

From the philosophical perspective it is possible to recognise a specific Hegelian cycle (Brandom, 2019). Basic on the dialectics mechanism (thesis, antithesis and synthesis) in the XX century there was possible to see the clash between thesis (social engineering) and antithesis by questioning them and pointing to various conditions of behaviour.

We are able to observe a specific Hegelian change in values from rationalism, social responsibility to individualism or biological determinism (Habermas, 1992). In the XX century we are witnessing the struggle between the social thesis and the biological antithesis. Now, the neuroscience helps to understand why individuals are obedient, behave in the same manner, use violence instead of taking rational action and we also have received methods that enable treatment based on genes and unconscious information (cyber neuroscience). Conversely, positive psychology supports the need to humanise the world. As a result of the process of humanisation of the world, security sciences are emerging. According to Hegelian cycle theory in the XXI century we should observe the next stage of development - a great synthesis which should solve all problems with conflicts between different points of view (Nadoh, 2016). Of course it is very idealistic expectation.

But if we take the lenses of idealism we receive the question of the principle of that synthesis process. They are the classic moral questions about the autonomy, freedom of individuals and, group safety have appeared repeatedly.

5 Conclusions

The conclusions of this article which could be useful in field of security studies and sciences are and for the practice in situations of escalating conflicts and threats (including natural disasters) are:

The action will be more effective and ethical if it is used to build defensive or negotiating strategies based on scientific knowledge and human rights (UN, 1946/1996, Boyd et al., 2014). These strategies should:

- introduce the scientific and philosophical basis that security professionals need;
- use the results of scientific research to enhance effectiveness of actions and use systematic observation or action research to monitoring the results;
- consider three dimensions: individual mental capabilities (including individual objectives and perception of reality), social (organisational and legal), and material (facilities, financial, economic military means and security infrastructure);
- take under account the impact of the high technologies revolution which gives opportunity to integrate the processes in science, technology and production and have a great impact on the transformation of the goal and content of education.

Thus, the main task of the modern intervention and education to peace process is to stimulate the creative process and the formation of the secure environment (e.g. technological culture) of the individual. The requirements for variability and flexibility of the intervention and educational process imply a completely different technology, focused on creative search and the development of a holistic concept from the point of view of the methodology for the implementation of security activities (Farnicka, 2019). It becomes necessary to change the structural component of the organization of the security services processes—technology and content. The rationalization of intellectual activity presupposes the emergence of a new specialization of security people (staff, activists, professionals, scientists) from the role of a theoretician or reactive observer to the role of a designer or part of the environment.

Notes

(i) *Security studies* also known as international security studies is an academic sub-field within the wider discipline of international relations that examines organized violence, military conflict and national security (Williams, 2012). In some countries create the discipline called the *security sciences*. The differences between them lies in the fact that the second focus not only on the global or state-centric security but also on the human and individuals feeling of security. Another unique feature of the security sciences is their rigorous research methodology (Piwowarski, 2020).

(ii) In the experiment' procedure participants were asked to rate the length of three lines and answer which line matched the length of that on the presented card. The participants were asked to respond alone and in a group. The study revealed considerable individual variation among the participants: approximately 5% of the participants always conformed to the group's answer, and approximately 25% of the participants held their own judgment every time. The remaining participants (75%) succumbed to conformity in at least half of the trials.

However, approximately 95% of participants disagreed with the group at least once. Once experiment finished, majority of participants expressed the belief that the others had given wrong answers (Ash, 1956).

(iii) Remarkably, if at some point the teacher approached the experimenter with doubts about continuing the shocks, the experimenter replied:

1. Please continue.
2. The experiment requires you to continue.
3. Please continue, it is absolutely necessary.
4. You have no other choice, you must continue.

If the first command was not effective, the second was applied, subsequently the third and fourth. If the subject did not obey after the fourth command, the experiment was terminated. If the teacher obeyed and applied another shock after one of the commands, during the next hesitation or question, the experimenter was to utter another command from the first to the fourth. At some point, all participants (teachers) asked whether another electric shock should be applied before pressing the subsequent button. It was evident that teachers were struggling with learner's suffering. The experiment was continued despite experimenter's refusal to terminate the experiment (Milgram, 1974).

(iv) According to this point of view human beings have specific mechanisms for specific forms of violence (for example: against stepchildren- Cinderella effect, intra-group violence is lower in individuals living in smaller societies,individuals may have a strong tendency to differ between in- group and out- group, which affects altruistic and aggressive behaviour, there is also evidence that both intra-group and inter-group violence is used in order to achieve goals in or outside the societies).

References

Abramova, M. A., & Krasheninnikov, V. V. (2018). High technologies as a sociocultural determinant of transformation of society. *Sibirskii filosofskii zhurnal, 16*(1), 91–101. https://doi.org/10.25205/2541-7517-2018-16-1-91-101

Alpert, E., Hayes, A. M., Yasinski, C., Webb, C., & Deblinger, E. (2021). Processes of change in trauma-focused cognitive behavioral therapy for youths: an approach informed by emotional processing theory. *Clinical Psychological Science, 9*(2), 270–283.

Appadurai, A. (2013). *The Future as Cultural Fact: Essays on the Global Condition.* Verso.

Asch, S. E. (1956). Studies of independence and conformity: I. A minority of one against a unanimous majority. *Psychological monographs: General and applied, 70(9)*, 1–70.

Ba, I., & Bhopal, R. S. (2017). Physical, mental and social consequences in civilians who have experienced war-related sexual violence: a systematic review (1981–2014). *Public Health, 142*, 121–135. https://doi.org/10.1016/j.puhe.2016.07.019

Bandura, A., Underwood, B., & Fromson, M. E. (1975). Disinhibition of aggression through diffusion of responsibility and dehumanization of victims. *Journal of Personality and Social Psychology, 9*, 253–269.

Baudrillard, J. (1992). *The Illusion of the end*. Cambridge University Press.

Blaustein, M., & Kinniburgh, K. (2012). *Treating traumatic stress in children and adolescents: How to foster resilience through attachment, self—regulation and competency*. Guilford.

Bond, R. M., & Bushman, B. J. (2017). The contagious spread of violence through social networks in U.S. adolescents. *American Journal of Public Health, 107*(2), 288–294. https://doi.org/10.2105/AJPH.2016.303550

Boyd, J. W., LoCicero, A., Malowney, M., Aldis, R., & Marlin, R. P. (2014). Failing ethics 101: Psychologists, the U.S. Military establishment, and human rights. *International Journal of Health Services, 44*(3), 615–625. https://doi.org/10.2190/HS.44.3.J

Brandom, R. (2019). *A spirit of trust: A reading of hegel's phenomenology*. Harvard University Press.

Brännström, L., Kaunitz, C., Andershed, A.-K., South, S., & Smedslund, G. (2016). Aggression replacement training (ART) for reducing antisocial behavior in adolescents and adults: A systematic review. *Aggression and Violent Behavior, 27*, 30–41. https://doi.org/10.1016/j.avb.2016.02.006

Buchanan, S. A., & Erdelez, S. (2019). Information encountering in the humanities: Embeddedness, temporality, and altruism. *Proceedings of the Association for Information Science and Technology, 56*, 32–42. https://doi.org/10.1002/pra2.58

Bushman, B. J., et al. (2018). Risk factors for youth violence: Youth violence commission, international society for research on aggression (ISRA). https://static1.squarespace.com/static/57530523f850829dde1dc031/t/5ad202726d2a73331c4449c1/1523712626745/isra-youth-violence-statement-2018.pdf. Access 10 Oct 2020

Cheung-Blunden, V., & Ju, J. (2016). Anxiety as a barrier to information processing in the event of a cyberattack. *Political Psychology, 37*(3), 387–400. https://doi.org/10.1111/pops.12264

Cheung-Blunden, V., Cropper, K., Panis, A., & Davis, K. (2019). Functional divergence of two threat-induced emotions: Fear-based versus anxiety-based cybersecurity preferences. *Emotion, 19*(8), 1353–1365. https://doi.org/10.1037/emo0000508

Cheung-Blunden, V., et al. (2021). Foreign disinformation operation's affective engagement: Valence versus discrete emotions as drivers of tweet popularity. *Analyses of Social Issues and Public Policy*, 1–18. https://doi.org/10.1111/asap.12262

Damasio, A. (2000). *The feeling of what happens: Body and emotion in the making of consciousness*. Harcourt Brace & Company.

Davidson, R., & Begley, S. (2012). *The emotional life of your brain: How its unique patterns affect the way you think, feel and live*. Hachette.

Dillon, K. P., & Bushman, B. J. (2017). Effects of exposure to gun violence in movies on children's interest in real guns. *JAMA Pediatrics, 171*(11), 1057–1062. https://doi.org/10.1001/jamapediatrics.2017.2229

Edmunds, J., & Turner, B.S. (2005). Global generations: social change in the twentieth century. *The British Journal of Sociology, 56*, 559–577. https://doi.org/10.1111/j.1468-4446.2005.00083.x

Eisenberg, N., Spinrad, T. L., & Eggum, N. D. (2010). Emotion-related self-regulation and its relation to children's maladjustment. *Annual Review of Clinical Psychology, 6*, 495–525. https://doi.org/10.1146/annurev.clinpsy.121208.131208

Farnicka M. (2017). Impact of cyberspace on individual safety and group security—a human developmental psychology approach. In: Ramírez J., García-Segura L. (eds) *Cyberspace. Advanced Sciences and Technologies for Security Applications*. Springer, Cham. https://doi.org/10.1007/978-3-319-54975-0_6

Farnicka, M. (2019). Action research as an element of integrative research model in social sciences. *Przegląd Badań Edukacyjnych (Educational Studies Review), 1*(28), 219–233. https://doi.org/10.12775/PBE.2019.012

Gabor M. (2011). *When the body says no: understanding the stress—disease connection*. Random House

Gruber, M. J., Gelman, B. D., & Ranganath, C. (2014). States of curiosity modulate hippocampus-dependent learning via the dopaminergic circuit. *Neuron, 84*(2), 486–496.

Guerrero, T. P., Fickel, J., Benhaiem, S., & Weyrich, A. (2020). Epigenomics and gene regulation in mammalian social systems. *Current Zoology, 66*(3), 307–319. https://doi.org/10.1093/cz/zoaa005

Habermas, J. (1992). *Postmetaphysical thinking: Philosophical Essays*. Cambridge University Press.

Haney, C., & Zimbardo, P. G. (1976). Social roles and role-playing: Observations from the Stanford prison study. In E. P. Hollander & R. G. Hunt (Eds.), *Current perspectives in social psychology* (4th ed., pp. 266–274). New York: Oxford University Press.

Jones, D. (2008). Human behaviour: Killer instincts. *Nature, 451*(7178), 512–515. https://doi.org/10.1038/451512a

Kandel, E. R. (2006). *In search of memory: The emergence of a new science of mind*. W.W. Norton.

Keyes, C. L. M., Waterman, M. B. (2003). Dimensions of well-being and mental health in adulthood. In: M. Bornstein, L. Davidson, C. L. M. Keyes, K. A. Moore (Eds). *Well-being: Positive development across the life course* (pp. 477–497). London: Lawrence Erlbaum Associates.

King, M. (1988). *How to make social crime prevention work: the french experience*. National Association for the Care and Resettlement of Offenders.

LeDoux, J. (2012). Rethinking the emotional brain. *Neuron, 73*(4), 653–676. https://doi.org/10.1016/j.neuron.2012.02.004

LoCicero, A., Marlin, R. P., Jull-Patterson, D., Sweeney, N. M., Gray, B. L., & Boyd, J. W. (2016). Enabling torture: APA, clinical psychology training and the failure to disobey. Peace and conflict. *Journal of Peace Psychology, 22*(4), 345–355. https://doi.org/10.1037/PAC0000213

LoCicero A. (2014). *Why "Good Kids" turn into deadly terrorists: deconstructing the accused boston marathon bombers and others like them*. Preager.

Lukas, J. (2013). *A short history of the twentieth century*. Belknap Press.

Maslow, A. H. (1966). The changing image of human nature the psychological aspect. *American Journal of Psychoanalysis, 26*, 148–157. https://doi.org/10.1007/BF01873431

Meeus, W., & Raaijmakers Q. (1986). Administrative obedience: Carrying out orders to use psychological-administrative violence. *European Journal of Social Psychology, 16*, 311–324.

Milgram, S. (1963). Behavioral study of obedience. *Journal of Abnormal and Social Psychology, 67*, 371–378.

Milgram, S. (1974). *Obedience to authority: An experimental view*. Harper & Row.

Nedoh, B. (Ed.). (2016). Lacan and Deleuze: A disjunctive synthesis. Edinburgh University Press. p. 193.

Nigg, J. T. (2017). Annual research review: On the relations among self-regulation, self-control, executive functioning, effortful control, cognitive control, impulsivity, risk-taking, and inhibition for developmental psychopathology. *Journal of Child Psychology and Psychiatry, 58*, 361–383. https://doi.org/10.1111/jcpp.12675

Perera, B., Faulk, C., Svoboda, L. K., Goodrich, J. M., & Dolinoy, D. C. (2020). The role of environmental exposures and the epigenome in health and disease. *Environmental and Molecular Mutagenesis, 61*(1), 176–192. https://doi.org/10.1002/em.22311

Peyrefitte A. (Eds.)l (1977). *Rapport Du Comite D'Etudes sur la Violence, La Criminaite et La Deliquence Society*. Du Comite D'Etudes sur la Violence, La Criminaite et La Deliquence Society.

Pinker, S. (1997). *How the mind works* (pp. 44–49) W. W. Norton & Company.

Piwowarski J. (2020). *Nauki o bezpieczeństwie [Security sciences]*, Warszawa: Difin.

Seligman, M. E. P. (2011). *Flourish*. Free Press.

Seligman, M. E. P. (2003). Foreword: The past and future of positive psychology. In: C. L. M. Keyes & J. Haidt (Eds.). *Flourishing: Positive psychology and life well-lived* (XI-XX). Washington: American Psychological Association.

Seo, J. S., Mantas, I., Svenningsson, P., & Greengard, P. (2021). Ependymal cells-CSF flow regulates stress-induced depression. *Molecular Psychiatry*. https://doi.org/10.1038/s41380-021-01202-1re trieved20.08.2021

Seville Statement on Violence. (1990). *American Psychologist, 45*(10), 1167–1168. http://culture-of-peace.info/vita/1990/Seville.pdf. Access 05 Sept (2018)

Sherif, M. (1936). A study of some social factors in perception. *Archives of Psychology, 27*(187), 60.

Sherif, M. (1956). Experiments in group conflict. *Scientific American, 195*(5), 54–59.

Trzebińska, E. (2008). Psychologia pozytywna. Wydawnictwa Akademickie i Profesjonalne.

United Nation (1946/1996). *Universal Declaration of Human Rights*. United Nations.

Van der Kolk, B. (2014). *The body Keeps the score*. Random House.

Williams, P. (2012). *Security studies: An introduction*. Routledge.

Zajonc, R. B. (2000). Feeling and thinking: Closing the debate over the independence of affect. In J. P. Forgas (Ed.). *Feeling and thinking: The role of affect in social cognition* (pp. 31–58). Cambridge University Press.

Zeanah, Ch., & Sonuga-Barke, E. (2016). The effects of early trauma and deprivation on human development—from measuring cumulative risk to characterizing specific mechanisms, *The Journal of Child Psychology and Psychiatry*, 1099–1102. https://doi.org/10.1111/jcpp.12642.

Zimbardo, P. (2007). *The Lucifer effect: Understanding how good people turn evil*. Random House.

Zimbardo, P. G., Maslach, C., & Haney, C. (1999) Reflections on the Stanford Prison Experiment: Genesis, transformations, consequences z T. Blass, *Obedience to authority: Current Perspectives on the Milgram paradigm* (pp. 193–237). Mahwah, N.J.: Erlbaum.

Zimbardo, P. G. (1970). The human choice: Individuation, reason, and order versus deindividuation, impulse, and chaos. In W. J. Arnold & D. Levine (Eds.), *1969 Nebraska Symposium on Motivation* (237-307). Lincoln, NE: University of Nebraska Press.

Conflict and Territory: A Legal and Metalegal Approach. The Case of Spain

Juan Cayón Peña

Abstract The territory has traditionally been one of the main sources of conflict. Since the borders of the states were determined after the Second World War, some of the main problems for security and defense have been determined by the separatist claims of those territories integrated into states from which they aspire to separate. There are cases in which the foreign influence is clear, but others in which the nationalist sentiment explains the desire for disintegration, as is the case in Spain. The territorial configuration of the Spanish State in the 1978 Constitution was an original milestone in the history of constitutionalism at the international level. More than forty years have passed since the approval of the Constitution and there is growing separatist pressure that the State is confronting with sometimes necessarily exceptional legal means. This chapter will try to confront the separatist problem as a source of conflict through an in-depth analysis of the legal and ethical issues involved from a philosophical point of view.

Keywords Territory · Conflict · Political philosophy · Spanish constitution

1 Introduction and Context

1.1 Forms of Government and State

Let us start this section by making preliminary reflections and laying the foundations on some concepts we will deal with throughout these pages. In general, they have been indistinctly used with some frequency, being common to use these terms to a mode of interchangeable words that all refer to the issue of power within the political community, and more specifically to the way that power is exercised in the different existing conceptions. Suffice it as an example Serrano Gomez (1977) defines the political regime as "mode or system of government of a country," in confusion with

J. Cayón Peña (✉)
Nebrija University, Madrid, Spain
e-mail: jcayon@nebrija.es

J. Cayón Peña (ed.), *Security and Defence: Ethical and Legal Challenges in the Face of Current Conflicts*, Advanced Sciences and Technologies for Security Applications,
https://doi.org/10.1007/978-3-030-95939-5_10

the concept of a form of Government. The confusion between the terms that we now analyze has been even more significant with modern theories in political science. These theories have preferentially adopted the terminology form of State to refer to the classic typology of the concept of State based on the territorial base or other traditional criteria and when referring to its political content (democratic, social, legal); Even to refer the heading forms of State to the classic distinction between the different forms of Government. From this terminological chaos arises the need to specify some concepts to which we will constantly refer in this work, at least in the sense that we will use, to avoid the logical problems derived from this conceptual tower of babel before which we face.

We can define a form of Government as the concrete materialization of a particular regime applied to daily life to make concrete political decisions. The denomination form of Government is nevertheless comparable to another that has also handled Lagüens Marquesan and Garcia de Vercher (1965), which is the political form, nothing but the embodiment of the political idea in a particular organization, that is, of a regime in the form of Government.

As Deustch (1970) points out, "since politics is decision-making by public means, it is primarily concerned with government, that is, the direction and self-direction of the great human communities." This approach to the concept of politics seems correct; what we fully agree with regarding the quote above is the conception of Government as a task of political leadership that will develop in every human community.

Government usually connects with the art of running a ship in the frequent comparison of ruling with the person who steers the ship of the State. Even the very etymology of the term government indicates this to us. In classical Greece, the term *kybernetes* designate the driver or helmsman of a ship, and both the words government and its derivatives find their last root in the term mentioned above. The modern meanings of Government, more focused on social and information control, have the exact etymological and conceptual origin, even though cybernetics.

In this sense, the different government forms would not be different ways of directing the political community in decisions. In the same way that individual decisions can follow different criteria, the Government of the peoples can entrust to different guidelines and develop in different ways. These forms of Government, as we have already advanced, will obey a specific regime, a series of convictions and beliefs of the community, values, and institutions that, with different depths within the community, will inspire its political development and with it, the development of one form of Government or another.

Regarding the specific forms of Government, Bobbio (1992) has already pointed out that their typology is one of those recurring themes on which almost all political writers have spoken out, each proposing the classification they deem correct. Moreover, he even opts for the form of Government that convinces him the most to govern the political community in question. This double function of studies on forms of Government is commonplace; We find a purely descriptive aspect when the authors teach us the different forms offered to the community to govern its political destiny through the Government concentrated in more or fewer people. However, a more prescriptive work indicates which is the form of Government that, in each author's opinion, best suits the historical-sociological peculiarities of the society in which it operates.

This has not been the only classificatory criterion followed by the doctrine; Thus, for example, we can distinguish between different possible forms based on applying mixed criteria: not only the number of people who exercise political domination (to which, by the way, he says, different types of State correspond), which in his opinion leaves essential questions unanswered, to classify forms of Government in a constant fusion of territorial, economic or constitutional patterns of classification.

1.2 The Genesis Towards the Modern State from a Continental Point of View

We usually find explanations of cyclical nature to clarify the change from one form of Government to another, through the paragon, such as if it were an organic body, the degeneration of one for the establishment of the next. A political structure is susceptible to change as a purely organic matter. There are many forces of all kinds that operate within it and affect it from the outside. As we have already pointed out, the most frequent solution is to consider political changes as a regular succession of pure or just and corrupted forms that tirelessly take turns governing the political designs of a community. Regardless of how the evolution took place, namely, either reform of the way the Government is exercised by adapting it to the different pressures that come into play, or radical and generally violent rupture in the face of the real or fictitious injustices and iniquities incurred by the dominators. Ultimately the classical explanation has ceased to have practical validity since the philosophical-political discussion has de facto left aside the issue of the forms of Government to focus solely on democracy. In our days, there is only one political form, democracy in its current sense, that of the modern State.

It is precisely at this point where it is essential to approach the very concept of the State, defining it as the contemporary political community that, organized, develops from the Renaissance to the present day. The coexistence of several elements, the territory, the population, and a specific legal organization, integrate the modern State.

Regarding the forms of State, we also find a classificatory variety that makes agreement difficult. As a preliminary step, we will say that all of them have in common to refer to the same State, that is, to the modern State, but little else. From the origins of the modern State, the classification criterion had been the territorial organization combined with the distribution of power centers. In this sense, modern States are unitary, federal, and confederate. Based on the criteria, this classification was recently complemented by new types of states of a mixed nature, such as the regional or autonomous Spanish State contemplated in the 1978 Constitution.

It is always appropriate to return to relations between man and society to develop it appropriately for a correct definition. It is essential to delimitate a concept of the

State and its scope of action and refer to its territoriality. We have already highlighted how, in our conception, the society, the community, is for the man, for the citizen. Zuleta Puceiro (1981) argues, "this is not only a part of the social whole. It is in a certain order of things but is projected based on its ontological dignity and subsequent dignity, both natural and supernatural, towards a dimension that transcends society". From this perspective, we can adequately understand the traditional solution to the relationships between the individual and society.

From this perspective, it is possible to maintain without contradicting himself that the person is ordered to the whole as the imperfect to the perfect since it is in the whole where a man can develop inadequate conditions the potentialities that are innate to him. In the same line of argument, the political society or the State should be understood as a smaller organic group of companies, although hierarchy itself, in the function of his goods and equipped with the autonomy necessary to fulfill the purposes to them their own, thereby contributing to the common good. This approach is especially appropriate to our final goal. The different autonomies should be subordinated to the superior community (the Spanish State) that recognizes them and gives political life. They would only make sense insofar as the fulfillment of their purposes (proximity to the citizen or decentralized management) contributes to the common good, that is, the peaceful development of coexistence state policy as a whole. Hence, to the contrary, when they operate against the common good represented by the State, they are delegitimized because they are contrary to the purpose for which they were constitutionally created.

It seems clear that the State is presented to us as the contemporary political community formed by a group of people and intermediate societies, united by emotional, historical, and legal ties, to regulate their coexistence and achieve good superiority to all of them. However, they have particular purposes and autonomies, precisely recognized for contributing to this higher purpose. Connects this approach precisely with the principle of subsidiarity to the now we approach as the way they should coordinate interventions and areas of action of different people and intermediate bodies that make up the political community called State. Subsidiarity operates based on a previous response to the problem of the so-called principle of totality that explains social relations, the relations of the whole with the part. To give a satisfactory solution to the issue of subsidiarity, we must first face and adequately assume the problems involved in the question of the principle of totality. Society is a unitary form in which two terms in principle opposed by their nature, such as the one and the multiple, the singular or individualized and the plural, are carried out simultaneously without being confused or mutually suppressed or eliminated. The reason for dealing with the principle of totality first, before dealing with the principle of subsidiarity is not whimsical; The doctrine has highlighted two compelling reasons for acting in such a way: first, since both principles operate in the corporate sphere, since they are not conceivable referring to a personal individuality, and society needs a principle of unity for its subsistence; And direction (which is the principle of totality itself), we will necessarily have to face the study of it first; but also, secondly, when the authority respects what we will call the principle of subsidiarity, it does so in compliance with its mission, the Common Good, which necessarily implies the same sense of discipline and hierarchy, sense of integration of the parts into the whole, that is to say, sense of totality.

Political science has been historically based on a broader than a simple description of empirical reality whose starting point must be sought not in the positive but rather a deep metaphysical root. The order of being is the reason and foundation of things, and as regards social relations, these are based, as we have already explained, precisely in the very nature of things, since man is, by nature, a social being. In this sense, society is a reality resulting from updating the person's sociability. As such, it supposes the idea of an order and its notes: First, a plurality of ordered elements. Secondly, their diversity—which implies inequality in the qualitative order-, the basis of the order of priority and posteriority that characterizes every order. Third, an ordering principle. Fourthly, a particular convenience of the elements that enable their connection; finally, the elements' exact relation to each other is determined by the order. For this reason, in no case should it be understood, in the logic we hold, that society is simply an aggregate of individuals, nor is it defensible that man is naturally called to resist life in common.

These assumptions are no longer accepted in a new worldview that does not escape the way of understanding the relationships between the whole and the part, that is, the principle of totality. This new position means the whole exclusively as a unit. From the traditional prism active participation and constant of each of the components of social life in it, the new logic intended occurs simply to affirm as much as possible the independence of each of the components of said totality. This hodiern approach about territorial tensions forgets that the plurality of its members characterizes the society. At the same time, it attempts to unify everything in the interest of what becomes a new guiding principle of totality but localized on one of the regional territories.

It is interesting now to address another side of the problem. The dominant liberal perspective, defended by Bentham or Mill, also introduces an essential qualification, a second element that Zuleta Puceiro (1987) summarizes in that "institutions, codes or constitutions are nothing other than freely agreed rules of the game." "The only foundation of authority is, precisely, the consensus about the validity of the rules of the game - the agreement on the agreements - and the need for it to operate as an instance of guarantee of the free play of interests." Simultaneously, advocates "the postulate of the necessity and legality of full and comprehensive development of human power, with no limits other than those imposed by its nature." The subsequent evolution of these approaches is clear; "the idea loses sight of that image of an austere and protective gendarme, to embrace the illusion of a great moral cause, capable of redeeming men from their situation of inequality and oppression "so that" as the supreme form of rationality, the State becomes a sovereign dispenser of those fundamental meanings on which the existence of the social totality rests." The evolution experienced goes from understanding the State as a simple guarantee of coexistence to understanding it as unique legitimized to act as a promoter of the conditions that "must" exist in the society thus understood, which implies a change in the way of understanding the essence of said society, which begins to be assimilated into a kind of raw matter. That it needs to be rationally molded following the dictates of science and social technology, basing all of this on an erroneous idea of social progress. The last consequence of such a conception is that the new principle of totality that inspires nationalism, among others, is necessarily opposed to the social plurality that traditional thought understood to be consubstantial to the very concept of society. Totality would imply substantial unity, and consequently unity of the social body manifested in the highest possible degree of convergence or, if possible, unanimity, obviating all reference to what is proper to society according to the nature of things plurality and qualitative differentiation. In this new individualism, the concept of political society empties of any community connotation under the pretext of greater rationality, turning the State into a modern "objective" and autonomous instance guided by rational criteria in its action and that, consequently, is irrefutable in their decisions. The new pact is not destined to act as a regulating idea of the process by which some communities deposit certain powers in a common authority to precisely protecting the freedoms and powers that they do not transfer and retain for the exercise of their purposes. Pact is the hypothesis that explains the existence of the social totality starting from reducing man to the condition of the isolated individual concerning any form of intermediate sociability.

The founding pact is not the origin of the Government but of the social totality itself; the process of transmutation of individuals into a collective self-depositary of national sovereignty. The sole clause of the said pact can only be the total alienation of those who sign it to give rise to that new State that, in the nationalist case, it would be born of the independence that gives rise to the new sovereign power.

1.3 The Common Good as Justification of the State

We will now outline the question of the Common Good, understanding that it is an essential reference criterion to elaborate an adequate treatment of the subject at hand. We have already said about him that it is the end of politics. It appears to us as the meeting point between the principle of subsidiarity and that of totality; thirdly, that it is the final cause of the political order and, fourthly, that it must operate as a guiding criterion for the action of the social authority (of the State) and as the last justification for it.

With Widow (1984), we understand by well what is or can be palatable under any aspect, always insofar as it is perfection or a natural complement to the appealing subject. In a first approximation, we could approach the Common Good by saying that it is that good that belongs equally to the different members of society, the good of society itself. The Common Good is communicable, and that it does not belong to one subject to the exclusion of others, but instead is the good of the whole. The end transcends the parts since it is not proportional in particular to none of them. At the same time, it is the good of all to which it is communicated. Hence, all activity of the State, political and economic, is subject to the permanent realization of the Common Good, that is, obtain those external conditions necessary to all citizens to develop their qualities and their trades, of their material intellectual and moral life. This digression, more typical of the philosophy of law than of constitutional law, comes up to raise the issue of the Common Good in society, which must be understood in coherence with what has been sustained up to this moment, as an optional whole and not as an integral whole. Therefore, the Common Good is the natural *raison d'être* of society. Even though it is a good proper to the parties, it is not presented to us as a particular and exclusive good but is communicated to us all parties. Components of the social whole in a distributive way, thus giving with all intensity in each one of them without excluding the others. In this way, we can initially outline the concept of the Common Good of the social whole as the greater good of each of its parts, superior to any particular good that should be subordinated to the common, since only in this way does it achieve its complete condition of good.

Moreover, from this same conceptualization arises the main problem that the doctrine has raised when the Common Good and the private good come into conflict. The preponderant situation of the Common Good compared to the different particular goods lies in the fact that the difference between the Common Good and the particular goods is quantitative and qualitative because following it could not be concluded otherwise. If the end of the Common Good does, the community will be what it is. If the Common Good corresponds to an entity other than the sum of the individuals, it will not consist of the addition of individual goods but will be good with its content. Society is understood as something more than juxtaposed individualities. Common good must also be constituted as something qualitatively different from an addition of interests or particular goods. That does not mean that their participation in society is annulling the individual, but rather the opposite: the individual cannot be realized except in the community, that is, in a society directed by the authority whose ultimate goal is precisely the Common Good.

Given the above, the State is nothing but the historical incarnation in a specific time (the modern one) of a permanent reality in the different historical moments, and that is the political community. The State has not always existed, nor can we affirm with certainty that it will not cease to exist one day. As Del Vecchio (2020) points out, "(…) the word State does not denote a historical category of universal validity as the nineteenth-century state theory believed, but rather is a concrete historical concept. The State is not a constant and permanent social phenomenon, but a transitory historical form: transitory, but not in the sense of nineteenth-century anarchist interpretations, which prophesied the advent of a stage of humanity without political existence, but in the sense of something historically limited and unique".

Consequently, the State refers to a political community organized on a specific territory with some type of authority that governs; that was called in other historical periods "republic."

Having made this precision regarding the historicity of the State as the materialization of another superior and permanent concept in human history, we could argue that the concept of the State encompasses two different positions, although compatible: or the State is considered as a social structure or group, or as a more or less specific force or function, that is, either we understand it as a legally organized independent society, or we tend to identify the State with the capacity to act coercively since it is not in vain that the State reserves the legitimate use of force within social life and citizens only obey the behaviors ordered by state organs through force and even physical violence. In Spain, Santamaria de Paredes (1898) gives the term the definitive accolade when in his Political Law Course he points out that "observation shows us the idea of the State under two different aspects, depending on whether we consider this idea in its unity, or decomposing by the analysis in other ideas (those of end, means, and activity of the State)". "But the unitary idea of the State, is also conceived under two other aspects: in itself, as an abstract concept of reason; and in its historical manifestation, as it has been produced and produced in time, embodied in certain social organisms." Regarding the strictest sense of the term, this author highlights how the State is always related to a specific type of organization, organized society.

In our opinion, the State's primary function is the Common Good. Of course, it is crucial to create and maintain the law, but State does not appear for that. Instead, man uses both the State and the law to live in society peacefully and develop all the innate potentialities. Both the one and the other serve man to achieve his ends.

Sanchez Agesta (1990), for his part, highlights how the doctrinal positions around the definition of the State can be grouped into three main lines: firstly, we would find the deontological definitions, one of whose contemporary representatives would be the Frenchman Hauriou (1928), who understands the State as "a regime that adopts a nation through a legal and political centralization that is carried out by the action of a political power and the idea of res publica as a set of means that are put together to achieve the Common Good". Secondly, we also find other types of definitions with a rather sociological character in the style of Herman Heller (2011) or Weber (1984), who respectively understand the State as a "lastingly renewed dominance structure through a representatively updated common act, which ultimately orders instance the social acts on a determined territory"; and as "administrative legal order to which the work carried out according to the group by an administrative body is oriented and whose value is claimed not only for the members of the community, but for all actions that are carried out in the dominated territory." Finally, we could also find a series of legal definitions of the State, in a sense understood by Kelsen (1980) when considering the State as the totality of a legal order insofar as it constitutes a system that rests on a fundamental hypothetical norm.

Sanchez Agesta (1990) also points out that its functions rationally justify power. It provides society with the necessary means for its intellectual and cultural development and those necessary for its physical existence. To achieve these assets effectively, it will be necessary to coordinate the efforts of all political community members, distribute the burdens to be borne to achieve said assets, and guarantee the peaceful use and enjoyment of them by the members of the political community. This is the "integral good" or *bonum integralliter* that Saint Thomas has called the Common Good, originating a whole doctrine maintained for centuries. We found some precedents in Aristotle (2004): "if we observe that every city is a certain community and that every community adjusts for the sake of some good- because everyone does the things they do for the sake of what that they seem good-, it is seen that all their communities want some good, and very notably that one, which is the most important of all, and that it includes itself to all the others, it will seek the most important good of all. This, then, is the city and the civil community". What in our times we have come to call the State does not have as its purpose its survival as could be deduced from practice, nor the maintenance of the law. However, on the contrary, it must serve solely and exclusively the society that composes it, to the men who form it.

2 Why Do Societies Increasingly Delegate More Powers to Peripheral Entities?

2.1 *On the Way to Organize Power and the State*

Within constitutional dogmatics, the expression "form of State" has served to denote quite different realities. In some way, when Kelsen (1980) defined it by the legal mode of production resulting in the distinction between democracy and autocracy, he went back to a criterion Aristotelian according to which the political form refers to its essence, to the center of gravity of its power.

Another thing is that from its Greek source to the Viennese mouth, it has followed a bumpy course. Through Jellinek (1981) and the statist school, state forms swallowed up political forms. Because the State itself is a political form, and because sovereignty is that center of gravity, unknown on the other hand in pre-modern public law. Determining the form of the State involves determining who is the sovereign. After Kelsen, some Italian authors such as Biscaretti di Ruffia (1996) use the formula to signify how the State is structured in its entirety, giving the first in the classification between States of classical democracy, States social its's and authoritarian states.

Thus, if we took the expression form of State in this sense, we would be entering, without the slightest doubt in a controversial way, in the paths of political philosophy more than in those of pure constitutional law, through the contrast between "sovereignty" and " subsidiarity." Alternatively, the same goes to a remarkable rubric, forged by Gentile (2012) at the head of one of his books, between "political intelligence" and "reason of State." Because the expression form of State also knows a widespread use to apply to the organization of territorial distribution of power.

In this last sense, the State can be organized around a single center to establish sovereignty, which gives rise to the unitary State. Alternatively, it can arise from the pre-existence of various centers that form a unit—according to the constructive logic and non-destructive that is at the origin of federalism—, as in the Federal State. Finally, we can think of the preservation of the multiplicity of these centers, with no more integration than that allowed by international law, in which case we are describing the Confederation.

Nevertheless, there have been no more confederations since the American Civil War when the USA was born under the federal Constitution. The same evolution has the Helvetic; decades ago, it became federal despite maintaining the name. Only in international integration processes does the formula reappear in the collective imagination, precisely as opposed to the federation. There are the vicissitudes of the European Union to prove it.

In the same way, federalism is a jungle in which federalisms dubbed "dual" coexist with other so-called "cooperatives" and sometimes with some that we could call "pretended." Differences between the Federal State and the politically decentralized unitary State (what we call regional State) are minimal. Regional State, born in the heat of the Spanish republican Constitution of 1931, exported to the Italian one of 1947 and returned home in the current Spanish, maybe minimal by a kind of picturesque constitutional mimicry. However, despite the many anti-federalist reluctances that are nested in the States that were born unitary, regionalism pure has not been satisfactory. Italy, Belgium, Spain, or even in Britain, also always on the edge of not being State, but in this cultured stratum fiercely Unitarianism.

Schmitt (2011) pointed out that "an effective territorial decentralization is only possible today, in a pluralist party state, based on a federal organization (…); the federalism, i.e., a federal organization, provides, in such a democracy, the surest means to achieve territorial decentralization". Although perhaps, because nothing is silent, from the Hispanic point of view, and we have just touched on it in the previous lines, we should remember that in the federal State, the difficulties for a correct organization of the territory do not come from the "Federalism" but of "statism" D'ors (1989) Again, subsidiarity and sovereignty, political intelligence and reason of State.

If society is not, strictly speaking, but a society of societies, and if political society does nothing but crown civil society, the principle of subsidiarity guarantees freedom, a consequence of responsibility, while the principle of totality ensures unity. And authority. Thus, there is no division between freedom and power but rather a harmonious game determined by the pattern of the common good. Modern logic, on the contrary, that of contractualism, leaves no room for subsidiarity or the common good. In its robust version, it is the reason of State, which is imposed on the social disaggregation; or individualism, in its weak version, which dissolves law. For this reason, we maintain that most of the current discourses on subsidiarity move away from classical logic to settle in any of the modern versions and especially in the second, which is appropriately postmodern.

Thus, most of the discussions about federalism seem not to come from the logic of sovereignty. Alternatively, because they oppose the "larger" State, the "smaller" States, which can be more oppressive for the citizens, the closer they are, and which in themselves do not ensure decentralization. At the same time, they contribute to weakening old nations that guard an important moral patrimony, however much it is often squandered. Alternatively, in a paradox, they use subsidiarity to defend the States. In our view, only the social organicity allows an adequate territorial articulation, conjugating the *pietas patria* with functional regionalism.

2.2 *The Social Constructivism Typical of Current Constitutions*

The first constitutions that appeared in the modern State and mainly in the European continent fundamentally sought to organize the powers of the State starting from their separation. French revolutionary doctrines ended with the Ancienne Régime. They impose theoretical deconcentration of power that goes from being in a single hand to being divided between the legislative, the executive, and the judicial. Those powers are balanced against each other, avoiding, at least in the theoretical framework, the abuses of the past. However, the historical evolution of constitutions and constitutional law itself has left that first mission behind.

In our days, the Constitution is the highest legal norm and the master plan of the entire society. Constitution lays the foundation of what is politically correct, limiting the spontaneous social politicity to channel it towards the parameters considered as adequate, as tolerable by society, outside of whose limits nothing can be done, nothing should be thought about, nothing is possible. This is precisely how constitutions, the real ones or those that reside exclusively in the collective nationalist imagination, are conceived and designed. They are the culmination of a political power that transforms social reality to adapt it to a series of ideological principles, to keep it within a few channels or frameworks defined by the constituent power and outside of which the State tolerates no other civilized life. It is not exclusively about a transformation of the organization of power, which of course it is, but rather, as Sanchez Agesta (1990) emphasized in his day, "it penetrates the entire structure of the social order."

This primacy of the will of the constituent power, independent from history and frequently, even from social reality, like in Spain, has deconcentrated and decentralized power in different regional authorities. In 1978, there is no doubt that Spanish society did not require distribution of power throughout the national territory. On an equal footing between the different autonomous communities since the old foralism in Spain had been overtaken by Bourbon centralization, surviving only in some territories a constant social will or at least in the informal power structures of said territories, a solid will for more self-government, perhaps even in some sectors of full autonomy or independence. Constitution opens the door in a dirigiste way to what was in no way posed as a generalized social demand. It designs from the roots a new system of distribution of power with a territorial base. It offers the possibility of a territorial organization that breaks with the model of the last centuries of our history. Therefore, it should not surprise us that, following the same procedure, the nationalisms that consider themselves constituent and yearn for the birth of their States also aspire to configure society according to their criteria, independently even of the genuine demands of those who will be their own. Subjects set themselves up as architects of the new society that must adapt to the political postulates embodied in their texts and slogans. However, some are even legal norms.

2.3 Centrifugal Territorial Distribution of Power in the Current Political Architecture

The centrifugal movement that we can observe in the distribution of power in numerous cases in old Europe, or even in the American continent, has started to obviate the more classical doctrine that in previous pages we have pointed out concerning the common good, to the integration of what is individual in the plural, of the part in the whole. This traditional doctrinal corpus has been reserved for the highest shelves of libraries, those that only serve to accumulate dust.

On the contrary, abandoning these past mental schemes, current constitutionalism has been nourished by some factors that, although they have far greater significance than the territorial distribution of power itself, are also reflected in it. Of course, they will gladly be developed in the future when time and doctrinal judgment are more significant.

However, it deserves a brief reflection justification, to the less obvious, leading to different states to territorially distribute power. We will purposely state the theoretical reasons for such a decision: To surround the citizens with power; gain efficiency; respect historical differences and increase solidarity between the regions. These four purposes, jointly or separately, constantly appear as the justifying basis for the decision to recognize certain territorial powers that complement the power of the State, deriving to them some of the powers initially attributed to the State.

Unfortunately, comparing these motivations with the reality of things, it is reasonable to think that the difference between them could lead us to think about the failure of the objective sought. Focusing on the Spanish case, partitocracy has prevailed in regions where not only the same national parties are present as well as new ones that seek their differentiation precisely in localism, many times intrinsically nationalist and disruptive. Therefore, the same representativeness crisis suffered by the large national parties and unions affects the local level without bringing power closer to the citizenry, but rather by replicating the models intended to be overcome, although on a smaller territorial scale. Regarding the benefits in terms of efficiency, it is enough to look at the regional and national budgets to see how, in general, greater efficiency has not been achieved. However, quite the opposite, with a multiplication of current expenses and civil servant personnel that has led to that in some Spanish cities about 80% of the working mass is for the Administration in its different aspects. Increasing the deficit and, on too many occasions, becoming a source of corruption or socialization of incentives so that it does not compensate even being employed, not to mention the differences between citizens of the same nationality or the excessive increase in tax pressure to pay for all this.

We cannot affirm that it has been successfully achieved regarding respect for historical differences. In some cases because those historical differences were simply non-existent. In others, because the differential treatment has always been seen as an unacceptable privilege by a large part of the citizenry and, to a large extent, has been a source of limitation of fundamental rights (as in the case of aggressive language policies in those territories with co-official languages). Finally, the purpose of promoting solidarity between the regions does not seem to have been reached either. It is enough to review the statements of politicians and citizens in bars or the media, with a constant comparison of who robs whom, who contributes more to the common funds, or who spends them without contributing anything to the common. Getting to consolidate some regional clichés that hardly contribute to the good of the State as a whole, instead, on the contrary, they serve to deepen the regional resentments between the different areas and of these to the State.

2.4 Personalism and Law

A final factor that could have weighed in the centrifugation towards the regions of political power that was once entirely in the hands of the State in the modern legal doctrine on human rights and its profound cultural impact on the European heritage. There is no doubt that Europe is the land of human rights; it is the legal-political environment in which man and his circumstances are legally more protected in their physical and moral integrity. After the Second World War, Europe is a benchmark worldwide and has served as an inspiration and reference, making its protection and guarantee perhaps the most widespread image of our legal system.

The rationalist philosophy and the personalism behind the ideology of human rights pivot on the central, nuclear role of the human person for any legal system, becoming, at least apparently, the foundation and reason for politics and law. At the early stages of this doctrinal evolution, human rights arise as a guarantee against the encroachments of power, against the risks arising from the disproportion between the power of the person and the State. By constitutionalizing human or fundamental rights, they achieve the highest possible level of protection against everyone. However, deliberately or not, the issue of human rights was called to even greater heights, because over time, they have transcended the law, becoming part of the ethical or moral, in the manner of a new religion typical of Europe, de-Christianized, covered philosophically by moralism and that today is dominant in the West. This philosophical approach makes the Constitution the highest norm of the legal system and the absolute moral referent of society, an ethical referent of its behavior.

Modern Western societies have thus become incessant claimants of new rights, which has led the doctrine to classify them into generations. The culture of our time is a culture of rights in that permanently, but often theoretical, they are conquering new areas of protection and guarantee. Moreover, it goes to the extreme that, with the consolidated and legally protected rights, there are often other supposed rights not yet consolidated but that aspire to be, even based on others that already are. The most obvious example that occurs to us of this situation that we are trying to describe is the confrontation between the right to private property, unanimously recognized in the West with the highest level of protection, and the incipient "right to occupy" those properties that are not in use or even those that are. In the courts, final resolutions usually protect the first against the second. However, it is no less accurate that the latter seeks protection and justification in the social function of property, the right to decent housing, or the inviolability of the home. Something similar has happened with the rights linked to gender ideology. However, in this second case, the process has already advanced in some legal systems such as the Spanish case until the full equality and full legalization of the social demands of those who came out of conventionalism in their day historical, social status regarding marriage, adoption, or family.

Finally, with all of the above, in what interests us concerning the centrifugal movement of displacement of political power, we arrive at the fact that the new territorial powers and the societies that nurture them also consider the legal evolution of their institutional framework to be natural. Consolidation of what they consider their rights following the examples set out above. Thus, the political community endowed with a regional power aspires to see the fundamental rights of its political conception enshrined in its highest standards (the statutes of autonomy in the Spanish case), transcending these standards beyond their strictly legal function and becoming a new ethical and moral reference. Of course, this new moral cannot, by its nature, know any contradiction in legal norms outside its territory, and that has not emanated from the regional society itself. From his point of view, any correction from outside the regional perspective is illegitimate interference by external powers, even if the courts deny them reason based on formal law. This nationalism is fundamentally sentiment.

3 The Crisis of the State?

In the following section, we will briefly deal with the aspects related to the functional or horizontal distribution of power that may have contributed to the acceleration of the centrifugal process of political power in the States that have sought a territorial distribution. Typical of the last third of the last century has coincided in time with the crisis of the State. After the unification of the old European kingdoms and principalities into superior units resulting in the political map of today's Europe, those states that did not opt for the federal configuration from the beginning eventually ended up distributing political power to a different extent among the territories that made it up. However, they did so at a historical moment in which globalism is growing due to the creation of the United Nations and, especially, of what we now know as the European Union.

The creation of supranational power structures has dramatically weakened the sovereignty of the nation-states by blurring or erasing borders, shifting the center of power towards supranational levels, and being accompanied by a generally open culture that makes the traditions more typical from nations suffer in favor of shared citizenship. There is no doubt that this weakening of the strength of the unitary State, also coinciding with the territorial distribution of power within its border limits, has turned out to be a dangerous combination. That could be ruining those States that were not constituted ab initio by federal aggregation of different smaller States (since the confederative model has practically disappeared from the political scene).

Political power is dispersed between the new supranational and regionalist tendencies, weakening the State in a kind of new "federalism." That is not a federation of States as defined by the canons of the political science but an amalgam of relationships and legal-institutional forms that may well end up being post-state. Sovereignty and territory give way to other interdependent and interrelated sharing power.

Since the middle of the last century, constitutional law and political science have emerged a new and highly relevant political operator: the Constitutional Court. Constitutional Court undoubtedly operates far beyond the Administration of justice, from the moment it has attributed the authentic interpretation of the constitutional text. Its direct intervention in matters of greater depth, in terms of the territorial distribution of power focusing on unconstitutionality remedies and constitutional conflicts of competences between the State and the different regional powers or between them. Yes, the Constitutional Courts have grown in prominence as the centrifugation of power from the State to the territories. The Spanish case is a great example. Constitutional Court had interventions of enormous relevance since both the most conflictive statutes of autonomy and those other decisions of the territories in nationalist hands that could have been systematically subjected to its interpretation, like any attempt against the unity of the State.

There is no doubt that, as the maximum guarantor of the Constitution, the power of the constitutional courts has made them a "superpower," maybe superior to the three classic powers in which Montesquieu deconcentrates power, as Ayuso Torres (1996) rightly sustain. For the same reason, and the difficulty of a politically considered election of its members, the constitutional courts have been transformed into a judicial body to adopt political decisions that of course, it seems that in a democracy they would have to be in the hands of the representatives of the national collective.

The juridification of politics, which inevitably occurs when constitutional courts are above the rest of political operators and overcome the classic division of powers, has fostered a kind of the absolute rule of law in the hands of its components to which no parcel of social life can escape. Their role in the territorial distribution of political power, being necessarily centralized, has undoubtedly aroused the zeal of nationalisms that, rightly or wrongly (we opted for the second of these possibilities) find in the representation of the State through of the Constitutional Court an immobile bloc that curtails any possibility of genuine self-government.

Members of our Constitutional Court, which otherwise often do not even have to come from the judicial career, make judgments exempt from partisanship or ideology and increase difficulty when, in addition, and for the sake of that "legal invasion" of politics, they are interposed as a dam. Insurmountable and responsible for decisions that an executive that was genuinely responsible for executing what the representatives of national sovereignty decided in legal norms could adopt. In this way, there has been a paradoxical reversal of the most reasonable reality: Constitutional Court is involved by Government or central power, not by separatist. Constitutional Court is used thoroughly in limiting the most unruly territorial powers that could be repressed with governmental measures, leaving the judgment of constitutionality ex-post and not ex-ante to them.

4 The Spanish Case

In the Spanish legal experience, the territorial issue has historically been approached with the reality of the "jurisdiction" of Foralism, which allowed a reasonably broad legal and political autonomy without diminishing the superior unity of Spain in that historical configuration that led to being Empire. In intellectual purity, this figure is a precocious maturation of the experience that we know with the name of subsidiarity. Nowadays, "Return of civil society" has not provided better results in achieving the community well. It tends to camouflage a set of lobbies and pressure groups that are guessed under the withdrawal of the State of large parcels that cannot be simply abandoned. Moreover, in the economic and social sphere, the subsidiary discourse today almost always conceals the reality of neoliberalism. It is used to weaken the State, which, despite its historicity in origin and development, many times currently guards the natural politics of man better than separatism, Europeanism, synarchy, or big money.

Having made this first observation unavoidable in the analysis of the Spanish constitutional reality, let us now place ourselves within what has been called the "State of the Autonomies." In general terms, the Spanish Constitution of 1978 contains a territorial design hybrid, or at least intermediate, between the politically decentralized unitary State and the federal State. In fact, Spain is a paradigmatic example of that "third way" intermediate between the most decentralized State of the unitary State and the federal (or confederal) State itself. Moreover, the text that culminated the transition from the Franco regime to modern democracy, as in so many things, purely leaves the question open. Article 2, included in the essential preliminary title, with the correlate of its aggravated rigidity in terms of Article 168, affirms, on the one hand, that the Constitution is based on "the indissoluble unity of the Spanish nation, common homeland and indivisible from all Spaniards," for another to consecrate the right to autonomy of the" nationalities and regions "that comprise it. Title VIII develops and thematizes Article 2 but does not create a "regional" State. More correctly describes one that could become "regional," since the principle of voluntariness—of access to autonomy—, together with those of equality and solidarity, was at the base of the model that the constitutional text contained.

This precision, necessary for the purposes we are interested in, is confirmed by the fact that Article 137, heading the title as mentioned above, determines that "the State is organized territorially in municipalities, provinces, and Autonomous Communities to be constituted." In this way, the new system devised by the constituent legislator is based on the concept of those indeterminate autonomous Communities, which can also be understood by "nationalities" and "regions," as well as being referred to other territories endowed with unique features, opening their training not mandatory to bordering provinces with common characteristics, island territories, provinces with the historical regional entity, regions from a single province, territories that do not exceed that of a province and to those others not integrated into the provincial division (articles 143 and 144). Even throughout the process of drafting the constitutional text when, in a contradictory way with the assumptions on which it rested, there is a generalization, homogenization, and acceleration of the system of territorial autonomies.

The autonomic pacts of 1981 are a true constitutional novation due to the substance of the matter and, deepening the path opened by the "pre-autonomous" regimes, the division into regions of the entire national territory. The division would be imposed even for those who, like the cases of León and Segovia, they would have rejected it if they had had the opportunity. It also extends Government and superior court of justice, which was only initially planned for the "historical" communities and those who had agreed to self-governed by the way called "fast" in Article 151 of the Constitution. The final result is an entirely artificial autonomous map, sometimes contrary to our history, with brand-new Communities whose existence sometimes seems to be due only to ignorance, picturesqueness, chance, or even the design of dividing Castilla and not consolidating León. Imposing a uniformity in maximums (colloquially known in Spanish as "coffee for all" and in English "one size fits all") in the long run has created an ineffective and expensive system that has not served to satisfy nationalist claims, today more vigorous and challenging than ever.

Furthermore, *stricto* sensu, regionalization cannot be based exclusively on demands of administrative rationality, paradoxically, on the same argument as the one used in the past for centralization. However, its nature as a projection of an abstract nature right should not be exacerbated either. Certain self-government must imply efficiency, but to a large extent, it is required by freedom, which is applicable at the different levels of territorial organization. Nevertheless, of course, the difficulty lies in determining that specific measure, that proper term, which in a modern State quickly deviates towards claiming a federal organizational model incompatible with legal and historical reality or leading directly to secession via a supposed right of self-determination that is exercised against all rights supported by the longing for a past that never existed.

As a proposition, it seems that a system that combined a broad administrative decentralization for the common law territories would have endowed them with greater efficiency and proximity to the citizens. Reserve the recognition of certain political rights of self-government for historical reasons, thereby strengthening the differentiation that many lengthy for, would have been more reasonable from an organizational point of view and, undoubtedly, more convenient from a political angle. Obstacles and reluctance of all kinds could raise, but the disturbing litigation of a transference process never closed and in constant expansion is dangerous; the reality of current events that put the unity of Spain at serious risk; and finally, if any of the complaining nationalisms were to consummate the disconnection, more than likely, others would follow.

The progress towards greater cohesion that has taken place in recent decades due to globalization should not be forgotten, forging in our homeland a previously unknown cultural unity, which—despite the nationalist counter—has not stopped growing. We take it for sure that there will be those who deplore the sign of this trend. However, for the non-judgmental observer, the mystery of that tightest forge between the different Spanish territories in the cultural field will not be able to go unnoticed. Only the strict linguistic and educational policy in the hands of some autonomous communities oriented to erase Spain, their name, history, and reality, have been able to limit such cultural homogenization that, for instance, could be considered impoverishing. Will we oppose these centrifugal procedures, an attractive pedagogy of what our Spain has been and could continue to be so rich in its plurality of regional characteristics?

5 Conclusions and Learned Lessons

1. The territory has always been a source of possible conflicts of all kinds. Social life unfolds territorially, and man has always been a territorial animal. We all consider ourselves attached to a territory with which we maintain a material relationship because it nourishes us and provides the necessary resources for life, but also, and most importantly, an emotional bond. That is why historically, patriotism has been an essential element of territorial cohesion that has its roots in the depths of the human being.
2. Frequently communities try to expand their territorial sphere. Others have fought for the defense of their territory. The territory has been and continues to be in the focus of most of the fierce conflicts, for example, in the case of the second world war in historical terms or the current conflict in Crimea.
3. In postmodernity, the international community recognizes the borders between states, and with some exceptions, the external borders are peaceful. However, the postmodern context does see the territorial problems derived from internal territorial disintegration grow.
4. A romantic and sentimental perspective that does not respect historical truth is attractive and popular in territorial politics. If it is maintained and cultivated for generations, it becomes dangerous for the territorial integrity of states.
5. In order to preserve their territory, postmodern states should respect the classic principle of subsidiarity, rigorously exercising their unifying mission. Integrating diversity into the whole is the only authentic guarantee of the territorial survival of the states.
6. Given the territorial problems, political representation, and institutional comfort that states suffer in postmodernity, perhaps we should appreciate that the state formula is obsolete, and a new configuration of the political community is necessary. The State as we know it did not always exist and may not exist forever.

References

Aristotle. (2004). "La política", TECNOS, Madrid.

Ayuso Torres, M. (1996). ¿Después del Leviathan? Sobre el Estado y su signo, Speiro, Madrid.

Biscaretti di Ruffia, P. (1996). Introducción al derecho constitucional comparado. Fondo de Cultura Económica de España, México.

Bobbio, N. (1992). La teoría de las formas de gobierno en la historia del pensamiento politico. Fondo de Cultura Económica, México, tercera reimpresión.

Del Vecchio, G. (2020). Teoría del estado. Olejnik ediciones, Santiago de Chile.

Deustch, K. W. (1970). Política y gobierno. Fondo de Cultura Económica, Madrid.

D'ors, A. (1989). Una introducción al estudio del derecho. Rialp, Madrid.

Gentile, F. (2012). Inteligencia política y razón de Estado. en *Verbo* (Madrid) pp. 501–502.

Haouriou, A. (1928). Principios de Derecho público y constitucional. Reus, Barcelona.

Heller, H. (2011). Teoría del Estado. Fondo de Cultura Económica de España.

Jellinek, G. (1981). Teoría general del Estado. Traducción de Fernando de los Ríos, Albatros, Buenos Aires.

Kelsen, H. (1980). Compendio de Teoría general del Estado. México, Editora Nacional.

Lagüens Marquesan, G., & Garcia de Vercher, R. (1965). Teorías y formas políticas. Edix S.A., Madrid.

Sanchez Agesta, S. (1990) Principios de Teoría Política. EDERSA, Madrid.

Santa Maria de Paredes, V. (1898). Curso de Derecho Político según la Filosofía política moderna, la historia general de España y la legislación vigente. Establecimiento tipográfico de Ricardo Fé, Madrid.

Serrano Gomez. (1977). Diccionario de términos sociopolíticos. Everest, Madrid.

Schmitt, C. (2011). Teoría de la Constitución. Alianza Editorial, Madrid.

Weber, M. (1984). Economía y sociedad. Fondo de Cultura Económica, México.

Widow, J. A. (1984). El hombre, animal político; orden social, principios e ideologías. Academia Superior de Ciencias Pedagógicas de Santiago, Santiago de Chile

Zuleta Puceiro, E. (1987). Razón y totalidad; notas sobre la noción moderna de consenso social. en *Verbo* (Madrid), N° 197–198; y "El principio de subsidiariedad en relación con el principio de totalidad. La pauta del bien común", *Verbo*, (Madrid), N° 199–200, (1981)

Ethical Considerations in the Area of Operations. The Use of Unmanned Military Systems and the Introduction of Artificial Intelligence on the Battlefield

Juan Del Pozo Berenguer

Abstract Within the last several decades, military technology has been devising different systems which allow servicemen to perform certain tasks without having to place the human body into jeopardy. Explosive ordnance has been an area of special interest for unmanned systems specialists, but aerial craft is, perhaps, the discipline where evolution has been more prominent. However, there seems to be an ethical dimension to the use of unmanned systems, and establishing what is ethical and what is not poses some serious challenges in itself. We could argue that a State-backed policy is, almost by definition, an ethical one. But if we leave it as that, we would be justifying certain questionable policies. Something else is required.

Keywords Unmaned weapons · Artificial Intelligence · Operations · Warfare · Ethics

1 Introduction

For millennia it has been part of human nature to seek an alternative form of peace that could satisfy its promoter. Every century "*there seems to emerge a country with the power, the will, and the intellectual and moral impetus to shape the entire international system*" (Kissinger, 1994 p. 17). And naturally, this powerful drive has been at the root of man's resolve for self-sacrifice in reaching this goal. Sailors and soldiers alike have roamed the Earth in pursuit of glory, riches and causes, however sincere or insincere they may have been. And they have always been willing to make the ultimate sacrifice, a tough decision to make or, at the very least, follow when not fully committed to the ideals that led to them.

During the 1588 intended invasion of Britain by Phillip II's Great Armada,[i] the Spanish fleet had to contend with an unexpected threat. During the long and arduous voyage up the eastern English coast, having failed to rendezvous with the Spanish army in the Netherlands which would have been the landing force in England, they

J. Del Pozo Berenguer (✉)
c/Velacho Alto 16, 3° A, 28033 Madrid, Spain

J. Cayón Peña (ed.), *Security and Defence: Ethical and Legal Challenges in the Face of Current Conflicts*, Advanced Sciences and Technologies for Security Applications,
https://doi.org/10.1007/978-3-030-95939-5_11

faced the attacks of unmanned ships which had been set on fire and directed towards the Spanish men-o-war. Despite the fact that "*the fire ships (brulotes) had not started the conflagration they (British) had hoped for*" (Crowley, 2009 p.176), the mere fact of having to face a threat with no human enemies onboard to fight against was quite unsettling for the Spanish sailors. Perhaps if the technology used to steer the *brulotes* had been a little more advanced, few sailors would have made it back home.

But rather than the mere damage inflicted on the enemy, it has been the method employed that has evolved to a point where ethics has increased its share in the overall equation of warfare. Indeed one of the most dominant factors that began to influence the process of military thinking "*was the enormous increase in the cost of fighting systems of all kinds*" (Naval Operations Analysis 1972 p. 31). Especially if these systems have some degree of automation. The ever increasing advances in military technology and artificial intelligence continue to raise the question of whether there are limits to the implication of an actor in its pursuit of inflicting damage to its adversary while avoiding risk for himself.

2 What Is an Autonomous System?

In order to submit the use of unmanned vehicles to ethical scrutiny, it is imperative first trying to define what an unmanned vehicle is or, at the very least, what its use implies. Secondly, we must analyse its nature from three different perspectives: purpose, ability to perform its mission, and safety. These elements will come up as we go along.

In his extensive dissertation, Richard Wilson defines an unmanned system simply as "*an unmanned airborne vehicle without a pilot on board*". As a clarification for eventual discussion, it goes on by stating: "*A drone may be navigated by remote control or by onboard computers that give the drone a degree of autonomy*" (Wilson, 2014 p. 1). The definition, which may sound somewhat redundant and unprecise, is in fact very adequate. The key element in any unmanned vehicle, which we will refer to as a *drone*, is its level of automation. Or to put it more simply, the degree to which it can perform certain tasks without human intervention, e.g., from pure self-navigation, which has a very limited or no impact on the ethics dilemma, through automatically gathering information to be eventually used to carry out an attack, which starts to pose problems, to autonomously selecting the target and deciding to fire, which fully does get into ethical shaking grounds, whether the decision is based on previous clear-cut criteria (heuristic approach, see below) or on much fuzzier artificial intelligence (AI) algorithms. This is the element that will eventually determine the ethical acceptability of the drone; the higher the level of automation, the more systems and capabilities may be fitted into its command and control system, thus allowing for a greater variety of tasks, and the less need it has of human control in real time.

When evaluating a situation, we are "*focusing on various dimensions of the action. We can evaluate the person who is acting, the intention or motive of the person acting, the nature of the act itself, or the consequence*" (Furrow, 2005 p. 44). Finding a speeding car on the highway is a good example. Either a car is speeding or it's not. It only takes a doppler radar to track all cars and identify the one not abiding by the rules, prompting the system to snap a picture of the offender. The scope of the decision is so simple that it hardly requires a human being to process the information and act upon it because this action barely requires evaluation. Exceptions, or particular cases that should exempt the presumptive offender from sanction, can be evaluated later with human intervention (note how the possibility of rectifying the action alleviates the ethics problem).

In scenarios both more complex and impossible to rectify if an error has been committed, like the ones the military often encounter in operations, evaluating a situation demands a complex process of knowledge and experience. In the former case, the doppler basic algorithm barely has anything to evaluate. It certainly doesn't require human intervention. But other processes demand a higher degree of evaluation and replacing a human mind by an automated command and control system is much more of a challenge.

3 Risk Versus Effectiveness

The XX century was the real turning point towards a *safer* (for oneself) conduct of warfare, with the concept of unmanned vehicles, as technology has allowed, step by step, along with the introduction of far more advanced command and control systems, to enable these systems to perform a variety of complex tasks. World War II saw the birth of the first airborne missiles in history with the introduction of the V1, and then V2, designed by Wernher von Braun,[ii] later to be the father of NASA's rocket program. True, both missiles essentially followed a ballistic trajectory once the fuel cell was drained, but it wasn't long after the war that the first guiding systems were introduced to allow the missile to seek and engage a specific target, while manoeuvring to react to conditions that might have changed between launching and impact.[iii]

Also, the secular pursuit of weapons with longer range was but an attempt to strike at the enemy from outside their ability to respond, to reduce our vulnerability while increasing theirs.

Although this early use of unmanned missiles did not have much continuation until the XX century, other ways of avoiding danger to our troops while keeping or even increasing the ability to inflict losses to the enemy were assiduously experimented with. Colourful uniforms with tall *shakos*, designed to awe the enemy and facilitate recognition of fellow combatants, were progressively ditched in favour of protective helmets and simpler uniforms in subdued colours intended to make the soldier as invisible as possible, ending up (so far, as true invisibility is still being sought) with the so-called *camouflage* patterns, a perfect antithesis of the grenadiers' red jackets and tall bearskin hats or blue dolmans of old. Not without resistance from the more conservative of the military, it should be added: during the Boers War (1899–1902)[iv] the British soldiers were surprised by the Boers' style of fighting, which consisted, instead of standing up to face the enemy and advancing openly in the field, in hiding behind rocks and other parapets while waiting for the enemy to advance and come within range, and dispersing if overwhelmed to fight another day, a tactic—aided by their use of plain clothes instead of fancy uniforms—which was deemed very cowardly by the proudly traditional British of the time. This was expressed very clearly by no less than Lord Kitchener, who wrote: "*The Boers are not like the Sudanese who stood up to a fair fight...*" (Dixon, 1976, p. 54 ff.)

More significant in terms of protecting oneself from unknown threats, and more ideologically loaded, is the universal conscription system established by the French Revolution and eventually extended to most of the world ("the people in arms") now being replaced nearly everywhere by professional armed forces, in order to shift the war casualties away from the younger—and much loved, as they are the country's future—cohort of the general population to the far lesser numbers of professional volunteers, whose losses are received with much greater equanimity, not to say indifference, by the society. Time has unearthed Machiavelli's prophecy with regards to conscription that "*troops of this kind may be useful and good in themselves, but they are always dangerous for him who calls them to his aid*" (Machiavelli Ed 1997 p.52).

Reducing the danger to our own troops while increasing the enemy's exposure to our gunfire has been a constant effort throughout history. But unlike the evolution of traditional weapons systems like the cannon, we are now confronted with systems that actually have been fitted with the necessary software to allow them to operate with a high degree of automation, effectively finding a walk-around to the safety of the opponent.

Is there a correlation between the risk an individual has to undertake to compensate for the potential damage it can inflict? In other words, must a person have to undergo great risk to his own life as a price to pay for the opportunity to destroy an adversary? If so, how do we justify the development and use of unmanned systems for offensive missions? These, perhaps, are the considerations we must weigh.

Within the last several decades military technology has been devising different systems which allow servicemen to perform certain tasks without having to place themselves in harm's way. Again, airborne weapons have been the area that has benefited most from these developments. These developments have been aimed at securing a higher level of accuracy but, for some nations and organizations, a major challenge has been identified as critical: that these systems, while extremely difficult to develop, are done in such a way "*which would ensure that the technology is safe and acts like humans would*" (Dyndal et al., 2017 p.1).

However, establishing the ethical limits of offending the enemy from safety, particularly with the use of unmanned systems poses additional challenges. Risking—even the certainty of losing—your life for your country has been an important part of the military ethos since the remotest antiquity. Certainly, shields and other defensive equipment have always been part of the soldier's kit, but risk even if so diminished was always there, and killing enemies from a totally secure and sheltered position has never, until now, been feasible. Where is the limit of what is morally or ethically acceptable in the continuum that goes from a one-on-one combat with equal weapons, as it used to be done by certain African peoples, notably the Zulu prior to Shaka, to the contemporary soldier of an advanced nation that, from the comfort and security of a war room thousands of kilometres away coldly monitors how an autonomous machine selects a target and proceeds to kill it? The *sorites* paradox enunciated by Eubulides of Miletus, by which it is impossible to determine when a heap of sand from which individual grains are taken ceases to be a heap and starts to be a simple count of grains, is a good description of the implicit dilemma.

It could be argued that a State-backed policy is, almost by definition, an ethical one. But if we leave it at that, we would be justifying certain questionable policies, loosely encompassed by a cynical "the end justifies the means". Clearly something else is required.

In Herman Wouk's "*The Cain Mutiny*", while reprimanding one of his officers to impress on him that everything must go by the book, Lieutenant Commander Queeg explained rather incorrectly that there were four ways of doing things on board his ship: "*the good way, the wrong way, the Navy way and my way: I want things on this ship done my way*" (Wouk, 1951, p. 140). Even if differently worded, this way of thinking is relatively common in the military. What Queeg (and others of the same mind) failed to understand is that no two situations are identical, and because of this, it is extremely difficult for any automated system to adequately deal with a scenario no one has ever thought about.

An automated system may, more or less, evaluate such a scenario, but both the intention of the system's reaction to this situation and its ultimate consequence must be laid upon a person, and not an automated system. If, as an example, a combat system operates in automatic mode as programmed but the programmer has somehow, either by omission prompted by programme's economy or by mistake, failed to contemplate a certain situation, thereby forcing the system to engage a wrong target, we can hardly place the blame on the system. After all, the system may have performed brilliantly. Alas, the instructions on how and when to engage could have been erroneously fed into the system, with important consequences.

Ralph Waldo Emerson[v] had once stated that "*cause and effect, means and ends, seed and fruit cannot be severed; for the effect already blooms in the cause, the end pre-exists in the means, the fruit in the seed*". When considering ethical aspects, these must be measured with regards to relations between individuals and the environment they evolve in and must have a purpose.

Perhaps the *purpose* is the guiding aspect of ethics as, in a way, it determines the means by which it is achieved. In other words, by defining the purpose of a system, we are in fact determining its means to achieve it, and to an extent the very ethics of its ultimate goal. The means is precisely what leads us to consider where unmanned systems fit in the realm of military operations as an ethical tool, under the safe assumption that the use of force may be a legitimate course of action.

The mere existence of these vehicles already poses a certain deterrent capability, but unlike nuclear weapons (of which the ethics are related to the "superfluous injury or unnecessary suffering" criteria, see below) this one is relatively easy to develop and within everybody's budget. In a certain Machiavellian way, it seems that we are willing to embrace the possibility of achieving maximum damage capability at no cost in terms of our own lives but, interestingly, also reducing collateral damage to a usually very high degree of proficiency, a luxury the nuclear weapon, for instance, lacks completely.

4 The Legal Dimension

To solve the dilemma of the use of automated systems in warfare, we could resort to International Law (i.e., The Hague Convention, the International Committee of the Red Cross ICRC, and others). Unfortunately they regulate the use of weapons basically under the only criteria of whether they may cause what has been generally and consensually described as "superfluous injury or unnecessary suffering" (Hague Convention on Respecting the Laws and Customs of War On Land, Chapter I, Art 23:"*...it is especially forbidden...To employ arms, projectiles, or material calculated to cause unnecessary suffering...*; ICRC, Rule 70: *The use of means and methods of warfare which are of a nature to cause superfluous injury or unnecessary suffering is prohibited;* and similar terms in many other Conventions and Treaties).

This is part of the reason why landmines are subject to more precise and meticulous regulation than naval mines (to the extent that for these last there is not even an agreed definition)[vi] as the former are directed to individuals and because statistics show that about 80% of casualties from landmines are civilians, versus the conveyances (ships) targeted by naval mines, which are deemed to afford some protection to people onboard, and in any case these people are professionals aware of the risks. Therefore, international Law is of little use to solve the problem of the ethics of autonomous weapons, beyond the obviousness that autonomous weapons must comply with the rules laid out by International Humanitarian Law.

5 Offensive Versus Defensive

When considering the impact of unmanned systems, there is another aspect, besides the level of automation, that must be taken into account, namely the offensive/defensive nature of the vehicle.

Naval mine warfare is a unique case within unmanned systems as it essentially encompasses most possible scenarios and, because of its potential destructive capability, makes an excellent case study. For this reason, we shall take a closer look into its particularities and perhaps then we shall be in a position to understand some of the ethical dilemmas involved in the use of drones or unmanned systems.

Since naval mines made the debut in the American War of Secession,[vii] the evolution of these weapons has been astonishing. During the early stages of this device, the key to success was to place the mines on the target's route, and the trigger mechanism in the mine itself, thus converting this "devilish" device into a fully autonomous weapon. This was heavily criticised at the time, as it was considered very "ungentlemanly" to kill without human intervention and correlative exposure. While this debate on the ethics of mining raged on, ways to circumvent, trick, or more generally, de-mine, were been found and applied, while the mines themselves also tried to overcome those defensive actions, with the result that today's mines have reached an impressive level of automation, being able to detect what kind of vessel lies above and remaining dormant for very extended periods of time, waiting for the right moment to act, all without outside influence once they are laid. The naval mine can, in fact, analyse the environment and the traffic above, deciding which contact is a target, and setting priorities among them.

De-mining missions are mostly aimed at self-protection, but in some cases they have a clear offensive dimension (for instance when clearing a defensive minefield in preparation for an amphibious assault) not only depriving the individual of the risks involved but also increasing the offensive capability to an extreme that a person simply cannot match and for a virtually unlimited time. These systems have set the grounds for maximum power delivery with minimum safety concerns for the operator or designer. In other words, the ultimate sacrifice seems no longer necessary to make the enemy more vulnerable, and herein lies the dilemma and the ethical dimension.

Whatever the adversary's intentions, de-mining a field designed to harm our traffic represents a clear case of self-protection, to which the use of unmanned systems poses no ethical considerations, as its ultimate goal is not to inflict damage to our adversary, but rather save lives. Interestingly, and continuing in the field of mine warfare, placing these deadly weapons is in itself a clear offensive activity. True, there is such thing as a defensive minefield—as previously quoted—in the realm of naval warfare with the sole purpose of deterrence, but there's still an offensive component to the activity, despite making public its deployment (as international law demands).

An unmanned aerial vehicle, however, can enjoy several degrees of human interaction, but as a general rule its deployment can be done from a very safe distance and with no risk to its operator. The level of automation, therefore, may have ethical implications for any unmanned system depending on what the automation component of the vehicle is designed to do: deploy, select a target or activate.

6 Heuristic Versus Self-learning

We now face a new reality in the way of artificial intelligence which in part of its design contemplates the systems' ability to learn based on experience. But can a system be designed to learn from previous experience? Can artificial intelligence allow for a system to evolve its capability to evaluate and make decisions without human intervention? Jeff Colombe categorizes artificial intelligence in two major types: *heuristic-based* and *statistical* or *evidence-based*.[viii] The first is perhaps the most primitive of the artificial intelligence systems.

Arthur Clarke and Stanley Kubrick portrayed masterfully this type of system in their *2001: A Space Odyssey* novel and movie, where the central computer *HAL-9000,* which had a protagonist role, was designed with this kind of algorithm (HAL is an acronym for "*Heuristically programmed ALgorithmic computer*"). With it, a system can evaluate and decide on the basis of problems that have been previously thought out by a programmer, making not perfect decisions, but very good ones and very quickly. With this algorithm there is no room for improvisation. The decision, therefore, in equal circumstances is always the same, as previously defined. However, even in these relatively simple circumstances, there were moral problems. *HAL-9000* killed most of the crew because it was faced with an unsolvable problem: it knew the real mission of the expedition but was under instructions to "*keep the crewmembers ignorant of it until close before the arrival to destination*" (Clarke, 1968 p. 230). The conflict between the obligation to share the truth with its human minders (as per the basic programme) and its concealment (upon Government's instructions) which increasingly crept into the interaction with the human crewmembers, finally made HAL to fall into what has been called a "Hofstadter-Moebius loop",[ix] becoming haywire and starting to commit deadly mistakes, first severing communications with Earth, then trying to cover every bad decision taken with another worse one, much like a clever but in the end unsuccessful human criminal. But HAL didn't have the capacity of introspection which would have helped it to solve the dilemma by evaluating the ethics of its instructions and the subsequent actions. This capacity is a crucial component of real human intelligence that a computer not only does not possess, but cannot possibly do.

It must be noted that this dichotomy between the, so to speak, hard-wired instructions and the eventual directions to be applied in a particular set of circumstances, although having stemmed from Arthur Clarke's fertile imagination, is a real possibility in any intelligent system. The *clausewitzian* "fog and friction" will take good care that nothing in the battlefield will resemble what the planner—or in this case the programmer—foresaw.[x]

The second (*evidence-based*) type however, is designed to actually "*improve the performance of a piece of software, based on evidence present in measurement data*" (Colombe, 2012 p. 1). What this means in reality is that, whereas with heuristic systems[xi] the programmer assumes responsibility for the outcome of a system's decision, the latter approach allows for the system to actually learn from experience. In this case two decisions made under the same set of circumstances may not be identical, as the second one might be influence by the learning that has happened in between. And this poses a critical ethical dilemma, as human intervention may not be fully involved in the system's evaluation and decision process. To that end, who assumes responsibility for the evaluation and action of the system.

There are, therefore, significant differences between heuristic and self-learning machines. At first sight the ethical consequences of either system should be different, as in the first case decisions are pre-ordained, while in the second they are unpredictable, as they depend upon the previous learning experience.

However, things are not that simple. Computer chess players—a field experimentally used to measure AI advancement, as within its strict rules machines can be pitted against extremely intelligent humans—exist of the two types. In 1997, Garry Kasparov, probably the best chess player ever, lost against *Deep Blue*, a powerful but purely heuristic machine whose play was devised by a group of chess players gathered by IBM. *Deep Blue* was able to evaluate 200 million chess positions per second. From that blow to human pride, competition started to be between different automatic chess players, rather than against humans, soon left behind in this superhuman championship. Until 2017, when *AlphaZero*, a deep neural machine devised not just to play *chess*, but also *go* and *shogi*, hence not specialised, soundly defeated the then machine champions, *Stockfish*, *Elmo*, and a previous weaker but more specialised version of *AlphaZero*.

What is significant here is that *AlphaZero* was not actually taught chess tactics, but instead was given 24 *h* of playing against itself, playing millions of matches. It evaluates just 80,000 positions per second (compare with Deep Blue's 200 million), but its strength lays in the ability to learn, and hence to quickly discard less promising avenues of play. No need of a coach team to think beforehand of all possibilities and encode the moves for each one, just a more human approach without the human emotions. Although it is difficult to envision the circumstances under which an autonomous weapons system can be subjected to training of the *AlphaZero's* kind, the implications for unmanned military systems are evident.

7 Conclusions

The meaning of all this is that in the end it is all about a machine blindly performing mathematical algorithms, whether hard-wired or resulting from a session of test-playing it is irrelevant. What is relevant in evaluating the ethics of the use of intelligent machines in warfare is therefore not how intelligent they are (whatever the scale used) or how they have acquired the intelligence, but the fact that they allow their operator to give simple instructions (versus physically driving the machine all the way to the target) to tailor the residing instructions in the machine's intelligence to a particular case, and to do it from the safety of perhaps thousands of kilometres away from where the action is. On the other hand, we should also add that, although not central to this debate, intelligent weapons are also very accurate as a result of the refined mechanisms they contain, with which collateral victims are minimised, thus intrinsically complying much better than non-intelligent weapons with the International Humanitarian Law's injunction that "superfluous injury or unnecessary suffering".

However, this appraisal is just provisional. When Strong Artificial Intelligence, as per IBM's definition,[xii] arrives to the armament industry, which would entail far more human-like abilities such as self-awareness and introspection, we will then be able to apply the very human measuring rules of ethics and moral to the machine's decisions. Until then, we are just dealing with another—albeit more sophisticated—version of the discussion of the ethics of naval mines, or even of the Boers's "cowardly" hiding behind rocks to avoid the enemy's fire.

Ethical and legal aspects must be met jointly when addressing autonomous systems, as the former can be a guiding principle for the latter. Although legislation has a tendency to drag behind the evolution of military materiel, it is the human mind ultimately responsible for its appropriate use.

Abiding by the law to any military organization is of paramount importance. Military technology develops at a rate that the law simply cannot keep the pace with. Ethics and law are, therefore, two concepts that cannot be fully understood without each other in the field of autonomous military systems.

As was the case with the introduction of nuclear weapons, new forms of inflicting damage in the field of operations must undergo close scrutiny on the impact inflicted by its legitimate use. Firing a cannonball at distance requires no more than perfecting aiming techniques. But the moment the cannonball can actually think for itself, or at least be programmed to a great deal of automation with no risk to its technical programmer (not necessarily a professional military man), this places the ethical dilemma at a completely new level of considerations, with significant legal consequences.

Despite the centuries that separate the brilliant strategist Sun-Tzu from automated systems, his words seem to be very fitting in this complex case when the theorist stated that "*the general who does not understand the perils experienced by the troops will not be able to employ them adequately*". Although these words were obviously not intended for the use of automated weapons, the suggestion that in absence of any peril for the attacker offensive actions may be ill-conceived is still valid today.

Notes

(i) Philip II's Armada, known as *Felicísima*, set sail from La Coruña on April 25th 1588 "*venturing into unknown seas*" (Howarth, 1981, preface) with the purpose of invading England as an expedition "*exclusively in the service of God (and) it was God's will that he (Philip II) should rescue the people of England from their heretic queen and bring them back to the true and only church*" (Crowley, 2009 p.14).

(ii) Unknowingly, von Braun had set the grounds for the inauguration of the space age following the end of World War II. The liquid propelled V2 rocket had the power to climb to an altitude of 50 miles before plummeting against its intended target, more than 120 miles away from its launch point. His design would soon to be employed for sending men to space.

(iii) The most important examples of unmanned systems are Unmanned Aircraft Vehicles (UAV), Surface Vehicles (USV), Ground Vehicles (UGV) and Underwater Vehicles (UUV).

(iv) With the origin of the war still under debate, the Boers War took place following the period of the Napoleonic Wars, between the British empire and the South African Republic and the Orange Free State. Despite British victory, their opponents inflicted heavy casualties on the British on account unusual fighting techniques.

v. Ralph Waldo Emerson was a XIX c. American philosopher who leaded the American transcendentalist movement, advocating the goodness of the human being, only to be corrupted by the individualism of modern society.

(vi) International Law Applicable to Naval Mines. Chatham House, Oct 2014: "*In contrast to landmines, which constitute a prohibited weapon for the majority of states by virtue of treaty law, states regard naval mines as a lawful weapon per se with their use regulated by Hague VIII and customary international humanitarian law (IHL). There is no international law definition of what constitutes a naval mine. However, NATO defines naval mines as 'an explosive device laid in the water, on the seabed or in the subsoil thereof, with the intention of damaging or sinking ships or of deterring shipping from entering an area'. In other words, naval mines are designed to destroy or damage ships although, more often, their use is intended to deny the enemy access to operationally significant sea areas*".

(vii) Admiral Farragut´s dramatic entrance into Mobile Bay in 1864 during the American War of Secession, revealed the enormous dangers of creeping into unexplored waters. Although Farragut made it through a barrier of over 180 Fretwell-Singer and Gabriel J. Rains type mines, the passage claimed several victims, despite the fact that Farragut´s men could "*see (Confederate Admiral Franklin Buchanan's) men distinctly when at work*" (Melia, 1991 p.8). His approach to dealing with the mines was successful and immortalized him as a brave hero, but his action is hardly a lesson for today on account of the evolution this weapon has undergone.

(viii) IBM describes two types of AI: Artificial Narrow Intelligence (ANI), and Strong Artificial Intelligence (SAI), which has self-aware consciousness and of which there are no practical examples in use today. IBM further differentiates between three levels of real world AI, each encompassing the next one: Artificial Intelligence, Machine Learning, and Deep Learning, which resides in machines that use neural networks. For the purpose of this discussion, which concerns the machine's ability to take decisions not necessarily hard-wired previously, we will refer to heuristic AI (roughly corresponding to general AI including Machine Learning) and self-learning, which would correspond to what IBM calls Deep Learning.

(ix) https://everything2.com/title/Hofstadter-Moebius+loop

(x) Particularly the first of the two elements, as Clausewitz equates "fog" with uncertainty ("war is the realm of uncertainty; three quarters of the factors on which action is based are wrapped in a fog of greater or lesser uncertainty"). This is particularly relevant in modern warfare, as the accelerated operational tempo often does not allow confirmation of situational data.

(xi) Heuristic systems have a basic standard programme and a number of shortcuts for specific situations (*heuristic addons*).

(xii) By Strong Artificial Intelligence, IBM refers to the level of artificial intelligence in which the system ceases to differ from the human mind. This level of intelligence requires, however, of constant input and experiences. IBM, however, admits that as of today, reaching this level of intelligence, while not impossible, is for the moment unattainable.

References

Clarke, A. (1968). *2010 Odyssey Two (A sequel to 2001 A Space Odyssey)*. Del Rey.
Colombe, J. (2012). *Heuristic and statistical artificial intelligence*. AAAS.
Crowley, R. 2008 (2009 ed). *Empires of the Sea*. Random Trade Paperbacks.
Dixon, N. (1976). *On the Psychology of Military Incompetence*. Pimlico Random House.
Dyndal, G. L., Bernsten, A., Redse-Johansen, S. (2017). *Autonomous military drones: No longer science fiction*. NATO Review.
Furrow, D. (2005). *Ethics: Key concepts in philosophy*. Continuum.
IBM Cloud Learning Hub. (1991). *What is Artificial Intelligence* (AI)?
Kissinger, H. (1994). *Diplomacy*. Simon and Schuster.
Machiavelli, N. 1532 (1997 ed), *The Prince*. Wordsworth Editions Limited.
Melia, T. M. (1991). *Damn the torpedoes: a short history of U.S. Naval Mine Countermeasures*. Naval Historical Center.
Naval Operations Analysis, 1968 (1972 ed), United States Naval Institute.
Wilson, R. L. (2014). *Ethical issues with the use of drone aircraft*. University of Maryland and Baltimore County.
Wouk, H. (1951). *The Caine Mutiny*. Back Bay, Little Brown and Company.

Challenges Facing the Response to the Covid 19 Pandemic. Individual Versus Global Responses. Ethical Issues

Jordi Regí Rodríguez

Abstract The problems we are facing due to the global Pandemic caused by the Virus SARS Cov2 and the following Covid 19 have not been equally treated in countries like China, the USA, and some countries of the European Union, and this has had clear consequences on the development of the illness. Neither the lockdowns nor the quarantine periods have been similar in these countries, and this has caused severe problems and conflicts that will be difficult to solve. Moreover, some countries have approached a global issue as an individual one and not a common one, and this has been dramatically reflected in the infection and death rates. To conclude, it has been made clear that the non-global management of vaccines has favored an enormous failure of humanity regarding this severe Pandemic. Also, problems with the ethical issues in the pandemic response will be analyzed.

Keywords COVID 19 · Global and national responses · World Health Organization · Ethical issues · Pandemic

J. Regí Rodríguez (✉)
Antonio de Nebrija University Madrid, Madrid, Spain
e-mail: jregi@nebrija.es

J. Cayón Peña (ed.), *Security and Defence: Ethical and Legal Challenges in the Face of Current Conflicts*, Advanced Sciences and Technologies for Security Applications,
https://doi.org/10.1007/978-3-030-95939-5_12

1 Introduction

COVID 19 Pandemic is without any kind of doubt the most devastating situation in this century with no equal and similar parameters in the history of humanity. The central Committee evaluating the response to the Pandemic was clear and, in the report, COVID 19 Make the Last Pandemic:

> The Pandemic is a sign of how vulnerable and fragile our world is. The virus has upended societies, but the world's population is in grave danger and exposed to deep inequalities. Division and inequality between and within countries have been exacerbated, and the impact has been severe on people who are already marginalized and disadvantaged. In less than a year and a half, COVID-19 has infected at least 150 million people and killed more than three million. It is the worst combined health and socioeconomic crisis in living memory, and a catastrophe at every level. (The Independent Panel for Pandemic Preparedness & Response, COVID 19 *Make the last Pandemic*, May 2021, 4)

In under three months from when SARS-CoV-2 was first identified as the cause of clusters of unusual pneumonia cases in Wuhan, China, COVID-19 had become a global pandemic threatening every country in the world. Although public health officials, infectious disease experts, and previous international commissions and reviews had warned of potential pandemics and urged robust preparations since the first outbreak of SARS, COVID-19 still took large parts of the world by surprise. It should not have been so.

Nations on our planet traditionally take different approaches to solve or attack the illness but without a global vision of the situation. However, during the last 18 months since the Pandemic was declared by the World Health Organization (WHO, 2020) on March 11th, 2020, only some days after the outbreak a Public Health Emergency of International Concern January 30th, 2020, with a clear message:

> In the past two weeks, the number of cases of COVID-19 outside China has increased 13-fold, and the number of affected countries has tripled. There are now more than 118,000 cases in 114 countries, and 4291 people have lost their lives. Thousands more are fighting for their lives in hospitals. In the days and weeks ahead, we expect to see the number of cases, deaths, and the number of affected countries climb even higher. WHO has been assessing this outbreak around the clock, and we are deeply concerned both by the alarming levels of spread and severity and the alarming levels of inaction. We have therefore assessed that COVID-19 can be characterized as a pandemic. (WHO, 2020)

Countries of our planet have clearly shown an individual approach to solve a problem that has caused 4,543,447 total deaths up until September 4th, 2021, with many consequences that we are still suffering now.

In this chapter, we will analyze the main aspects regarding the main challenges in different countries and regions of the world to try to solve the Pandemic with its main elements and different results as the Pandemic evolved since its declaration and the problems facing vaccination and the non-global scenario regarding distribution and side effects.

Ethical issues about pandemics and governments will also be reviewed because of the last events in some countries.

2 The Public Response to the Pandemic

SARS Cov2 is a virus of zoonotic origin "whose appearance was very probable" (The Independent Panel for Pandemic Preparedness & Response, 2021). They emphasize that detecting a new pathogen as soon as possible is essential to containing it.

The emergence of COVID-19 was characterized by a mixture of early and rapid actions, but also by delay, hesitation, and denial, with the result, that an outbreak became an epidemic, and an epidemic spread to pandemic proportions in a brief period.

The Pandemic has had a massive impact on our world with dramatic circumstances: More than 148 million people infected, and more than 3 million have died in 223 countries, territories, and areas. (The Independent Panel for Pandemic Preparedness & Response, COVID 19 *Make the* last Pandemic, May 2021, 4)

WHO learned of the outbreak on December 31st, 2019. A month later, on January 30th, 2020, it declared a public health emergency of international importance, which is its highest level of alert, following international legislation.

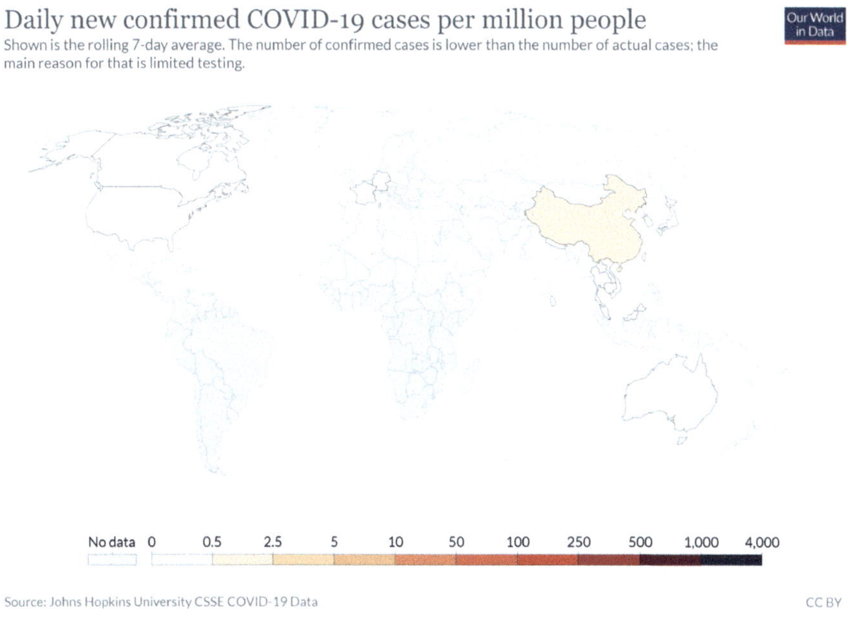

Cases on January 20th, 2020

Despite the multiple uncertainties that continue to exist at a biological, clinical, and epidemiological level concerning this new virus, what already seems clear is that each country has responded—or is responding—to the same threat with different measures and/or with different timing. This fact means that the epidemiological curves of the affected countries are behaving differently and that the social and economic costs of the respective responses may be different. But as it has been said, the Pandemic was declared two months later. (Centers for Disease Control and Prevention, 2020).

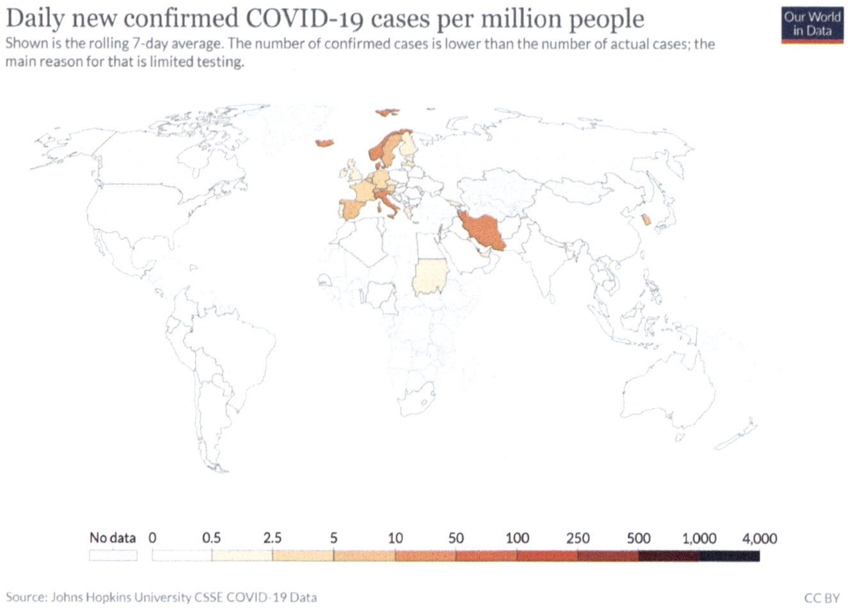

Cases on March 11th, 2020

The WHO response was quick and efficient, but different countries and areas of the world were not, and this had other effects on massive situations according to the decisions taken by the foreign governments.

> Governments have sought to contain the COVID-19 Pandemic by imposing restrictions on activities that enable SARS-CoV-2 to spread rapidly through large networks of people. Common measures have included travel restrictions, closure of schools and places of worship, and stay-at-home orders, although approaches and timetables have differed greatly. An important phase of crisis management is the one dedicated to the lessons learned. Managing the COVID-19 coronavirus pandemic encompasses many factors that are still in full evolution. (Lazarus et al., 2020, 1)

> These measures include school closings, travel restrictions, bans on public gatherings, emergency investments in healthcare facilities, new forms of social welfare provision, contact tracing and other interventions to contain the spread of the virus, augment health systems, and manage the economic consequences of these actions. (Hale et al., 2021)

The rapid spread of COVID-19 globally has been met with an extraordinary range of government responses.

Despite the multiple uncertainties that continue to exist with this virus at a biological, clinical, and epidemiological level, what already seems clear is that each country or region has responded to the same threat with different measures and with different timing. This fact means that the epidemiological curves of the affected countries are behaving differently and that the social and economic costs of the respective responses may be different.

But in the case of this, Pandemic was characterized by a mix of some early and rapid actions, but also by delay, hesitation, and denial, with the net result, that an outbreak became an epidemic and an epidemic spread to pandemic proportions.

> In countries where citizen surveillance and control is limited, the success of lockdown to reduce COVID-19 depends on a complicated voluntary process of information processing and institutional compliance. Specifically, individuals and communities need to trust and adhere to advise from scientists, politicians, and law enforcement, while ignoring disinformation and conspiracy theories. It is possible, however, that the Pandemic itself (and subsequent lockdown) not only relies on but may change the extent to which people trust institutions.... (Sibley et al. 2020, 2).

We will see later the analyze of the different countries approach to responding to the Pandemic how the situation has produced results with significant evidence that prove how a severe response produces better results, talking about the incidence of the Pandemic, number of deaths with the subsequent waves of the illness that reflects a more flexible behavior.

The robust containment has been dominant in the Asia and Pacific Countries. Still, in the United States of America and the European Union, the situation is entirely different. Even after one year, the data shown in this chart is clear about that.

This aggressive event affects the population with different attitudes according to the different national lockdowns that are less cohesive, and governing bodies are divided on the course of actions (Sibley et al. 2020, 2).

Social and economic problems arising from the COVID-19 response leave open a growing number of essential questions posed.

> Will increases in patriotism and governmental trust continue? Who will recover from loneliness and loss? Longitudinal work conducted after 9/11 in the United States found that the majority of people were resilient, but a sizable minority went on to develop post-traumatic stress disorder. (Bonanno et al., 2008, 665)

The following chart is comparing situations in the countries studied:

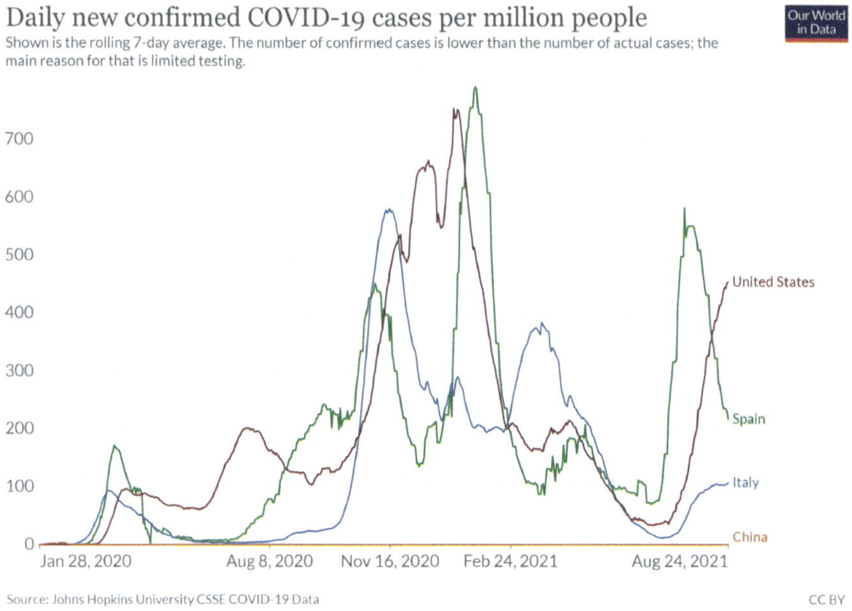

Covid 19 cases from January 28th, 2020/August 24th, 2021, USA, China, Italy, Spain

As we can see, cases in China are entirely irrelevant until now, but in the case of the USA, we can perfectly see the five waves. The fifth wave in the USA is probably produced by the people that are deniers with vaccines.

Probably the obscure management of China was compensated by an enormous activity, as we will see in the next chapter. Still, the confusion with the nonintervention of President Trump in the case of the USA and the denial of the existence of such a pandemic was a complete catastrophe in the resolution probably only compensated by the arrival of President Biden in November 2020.

If we return to Europe, Spain has five waves up until now, and in the case of Italy, there are four waves.

The first wave in Italy and Spain shows the effects of the lockdown. The second one is clear with the impact of a relaxed summer, continuing with the third in both cases produced by the Christmas holidays. Finally, the fifth wave in the case of Spain is produced by the total relaxation of the measures this summer.

Comparing countries death rates also shows the principal effects of the Pandemic where the impact of the vaccination is proven in the countries studied, even though there has been an impact of the Delta strain.

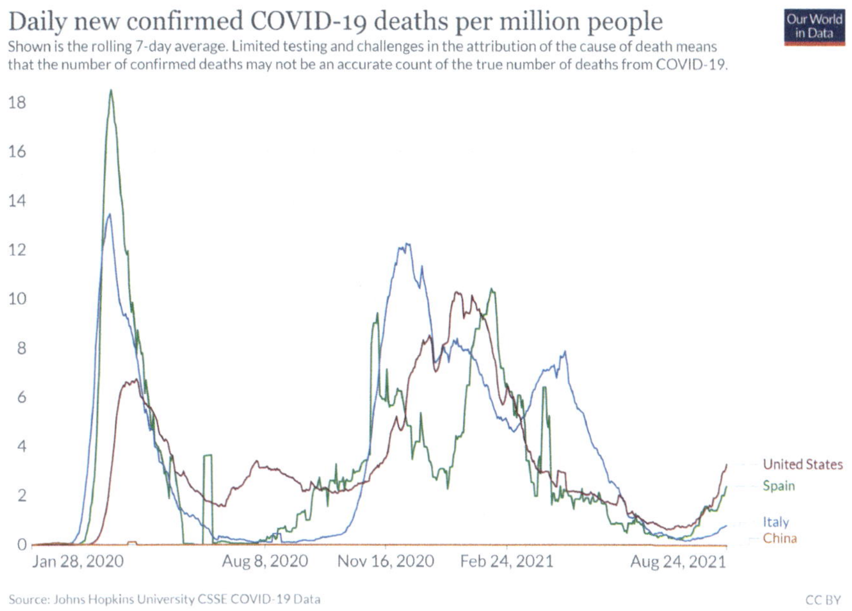

Total deaths per million people from January 28th, 2020/August 24th, 2021, IN USA, China, Italy, Spain

Finally, it will be interesting to look at the vaccination policy of the states studied in this article to compare them in Italy, Spain, USA and China.

Let's talk about the vaccination campaigns in the different countries. The situation has a clear impact in the civilized world with an apparent effect on countries like the USA or the UK with a significant difference in numbers of vaccines inoculated to the population.

China is again assertive in the situation and, of course, without several peaks as we have had in Europe.

After the resolution of the vaccine crisis and the problem with the doses, it is having a pretty good response but maybe a little bit too late.

The European Union showed a very efficient common approach with the approval of the EU Vaccines Strategy in June 2020 (European Commission, 2020). This strategy is undoubtedly one of the most important agreements with several advance purchase agreements and a global solidarity effort to make vaccination possible.

EU, in June 2021, the centralized distribution program, delivered more than 333,678,903 doses (European Center for disease prevention and control, 2021).

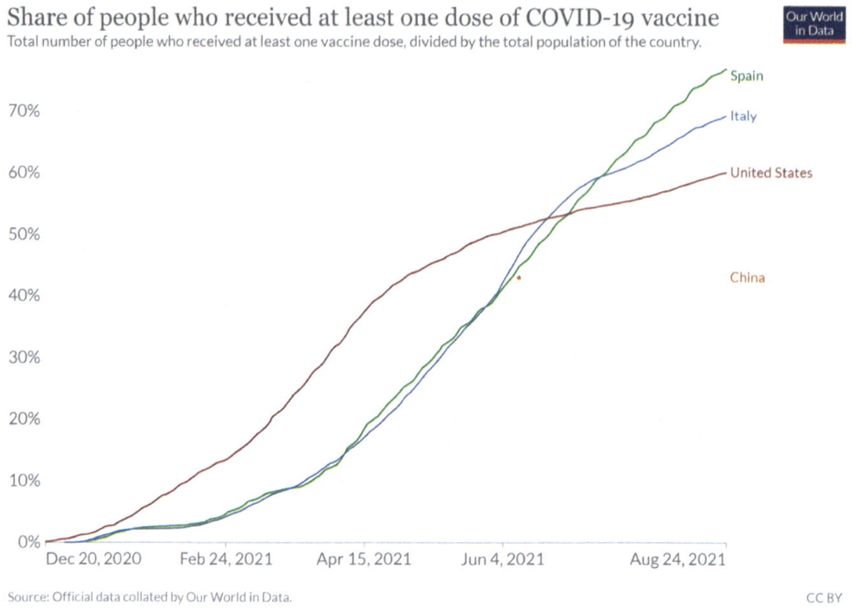

Vaccination people per million people from December 30th, 2020/August 24th, 2021, IN the USA, China, Italy, Spain

Spain is the leader of the vaccination in the states that we are comparing, and there is an apparent effect of the anti-vaxxer in the USA, where the curve is flattened at the beginning of summer.

We must indicate the slow start of the European Union countries with the distribution problem motivated by the problem with the Astra Zeneca Vaccine and the non-distribution of doses to the EU in the first months of 2021. After resolving the issue, a rising line seems to start flattening in Italy, but it is still too early to confirm it.

Global vaccination strategies have shown a more efficient impact in the developed world and can unify countries in production and distribution (Covid-19 Vaccines Global Access, COVAX, 2020).

The efforts done to produce this vaccine are shown in the following graph comparing a standard vaccine and the COVID-19 one.

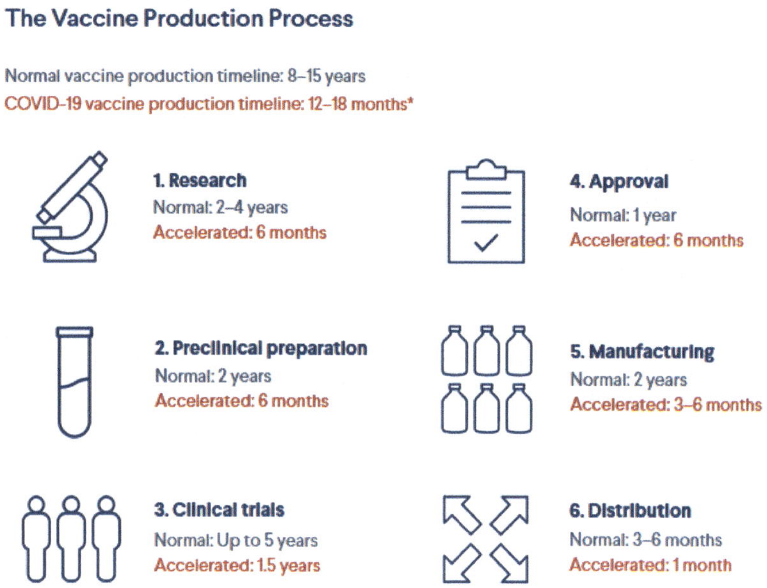

Source: Johns Hopkins University (2021)

Regular Vaccine Production Process compared COVID-19

As we can see, the vaccination production process is reduced from 8 to 15 years to 12 to 18 months, showing a massive effort in the industrialized world, but it is still too early to know if this campaign will end the pandemic (Council of Foreign Relations, 2021).

Vaccination in the developed world will be of no use if the disease is not contained in other countries. Therefore, it is necessary to vaccinate all the countries in a global health strategy that recognizes extensive immunization as a global public good.

But there are doubts within WHO about the solidarity of countries in the distribution:

> International collaboration among scientists was critical to vaccine development, but now weak cooperation between nations is a major barrier to achieving worldwide vaccination at the scale needed to end the pandemic. Vaccine equity isn't just a slogan; it protects people everywhere, protects the existing shots from new vaccine-resistant variants, and strengthens the international community's ability to stop COVID-19. (Ghebreyesus, 2021)

After more than one year and a half of Pandemic, all the situations are more precise. Still, it must be confirmed that it is not an easy problem nor a logical and easy problem to be solved, and we have to wait to see if the independent responses are better than the global ones even though the states adopt no clear global response on earth.

3 Different Responses to the Pandemic: China, Italy and Spain, and USA

There are different approaches to respond to the problem. In the case of the COVID 19 pandemic, all the situations that have to be solved were not easy because of the quick expansion of the illness from the end of December 2019 until the declaration of a pandemic by the WHO on March 11th, 2020.

> Every district, every county, every state could make decisions and keep them to themselves, and we just have uneven applications of public health recommendations in a way that I can't imagine any other country does. (Gandhi et al., 2020)

The most iconic responses to the corona-virus pandemic in developed Western countries have had more to do with individualism (expressed in the empty shelves of US supermarkets) and the measures that governments take in the face of a media society, and which has as its priority the typical western electorate that is manifested right in the US, rather than preventing the spread of the virus. Quite in contrast to China's overwhelming response (which, despite the initial stumble of taking a month to report the first outbreak, corrected the course in time).

Values of security and freedom are irreconcilable at the extremes. Each political system finds the balance that suits it or the one it can sustain, more space in democracies, more security in planned dictatorships. It is in emergencies that this tradeoff is stressed. Each country finds its balance between the benefits of mapping the route of the virus and imposing restrictive measures on the movement of people and goods, on the one hand, and, on the other, the invasion of privacy and the impact of the market economy a tremendously negative externality (Montani, 2020).

An essential phase of crisis management is the one dedicated to the lessons learned. Managing the COVID-19 coronavirus pandemic encompasses many factors that are still in entire evolution, so it is not possible to draw lessons for the future. Crisis management models consist of systems and procedures. The procedures—protocols in health terminology—comprise an evaluation of the general context of the crisis (risks and objectives), how it is intended manage (operational concept), and the measures taken to develop the above (action plans). The system—not the entire National Health System but the part that deals with crisis management—monitors the situation, proposes measures, and adopts or executes decisions related to it. The system works permanently between the Pandemic and Pandemic (inter-pandemic period) to monitor ongoing crises and develop preparedness and response plans for the following. When a new such as COVID-19 is activated to adapt previous strategies and plans to the specific characteristics of the new Pandemic, the management of contagious diseases has traditionally remained in the health field.

In contrast, infectious diseases were considered a health problem. However, as the epidemics have created damage collateral to those of health, it has become necessary to involve those responsible from other areas of the Administration. Moreover, the need has become peremptory with the emergence of pandemics, epidemics of global proportions, and disruptive effects that transcend health and affect societies' way of life and prosperity.

The response to a crisis like the COVID 19 must be treated from both a health and security perspective. (Arteaga, 2020).

3.1 China

In December 2019, several patients with pneumonia of unknown origin were admitted to hospitals in Wuhan, China. Later tests on a cohort of patients admitted between December 16th and January 2nd found 41 with COVID-19.

On December 24th, the Chinese health organization started to doubt these cases of pneumonia.

After Christmas in China, it was proved that existed a transmission of human to human.

The primary response from China was quick and effective:

Immediate lockdown of the city of Wuhan and other cities.

All the people in the region were forced into quarantine, restricted movement between and within cities, imposed severe restrictions on hundreds of millions of citizens, enforced the use of Health QR codes on mobile phones, carried out sterilization of buildings and streets several times a day, isolated all suspected cases in facilities, and increased hospital capacity by more than 50,000 beds to test, admit and treat all patients.

All these measures proved a quick response that dramatically reduces the transmission of the virus to other regions flattening the epidemiological curve.

This approach has helped to rapidly reduce pressure on the health system. These measures, however, have a high economic and social cost associated with the isolation of all positive cases in health facilities and the rigorous restriction of individual freedom of movement.

In addition, it requires up-to-date data at the population level and robust data management capabilities, something that many other countries lack. But it has been the most efficient response to the virus pandemic, although it is not easy to adopt it in Europe or the United States.

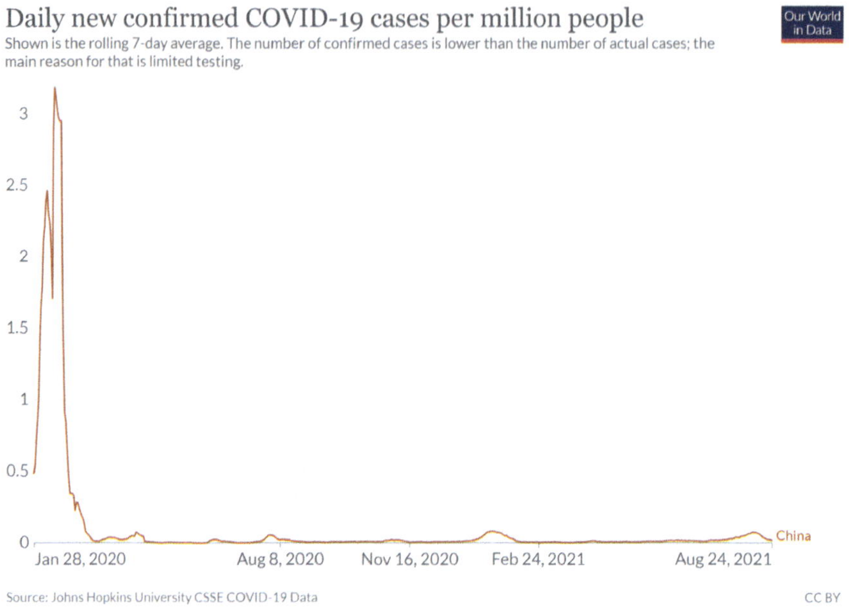

Daily new confirmed cases in China from February 5th, 2020/August 24th, 2021

The advantages of this severe system are a quick reduction of the pressure at the hospitals and, of course, essential cooperation among all the authorities. Still, there are also disadvantages because a necessary rise of the budget in health is needed to treat all the people and of course the control of the population in terms of security and data are critical showing possible attacks to fundamental rights of the people. It doesn't seem very easy to apply in other countries but China. (Chaccour, 2020).

As we saw in the previous chart, the problem in China seems to be solved by talking about the Pandemic.

If we talk about dead people, it is also clear that China has reached an outstanding result in a brief period, as we can see in the following chart:

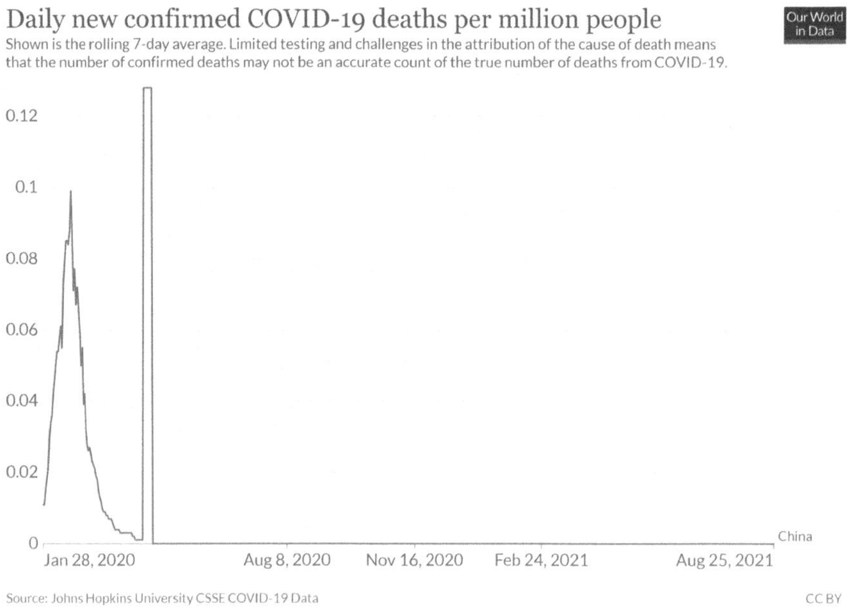

DEATHS in China from February 5th, 2020/August 25th, 2021

The results of these rapid and drastic measures are apparent. They helped slow the spread of the virus from Hubei to other provinces, which showed a less steep epidemiological curve. China reports about 20 new cases every day, up from 4000 patients a day during the epidemic's peak.

3.2 *Italy and Spain*

At the European level, health competencies fall fundamentally on the States, which has impeded effective coordination of crisis management in its most acute phase and now complicates a standard exit from confinements, with shared mechanisms of control and monitoring of infections. Without extensive coordination, in which Spain could use its experience with the disease, and the management of intense flows of travelers, disturbances in the internal market and the Schengen zone will slow down recovery in the economy. However, they will facilitate the reappearance of new outbreaks. For now, there are many doubts on immunity to COVID-19, but when more information becomes available, it may homologate a health certificate at the European level. In this way, they would combine health and mobility recovery criteria, avoiding quarantines generalized, which on the other hand, will require PCR or reliable rapid tests (García-Basteiro & Legido-Quigley, 2020).

Italy and Spain adopted similar measures with a similar result. The first closure of the provinces in Italy was announced in the north on March 7th (more than a month after the first cases were reported on January 29th, although the virus is believed to have circulated since mid-January), followed shortly by a nationwide shutdown.

The lockdown in Spain, which began on March 14th (a month and a half after the first cases were detected in the Canary Islands); It takes 5–7 days (the median incubation time) to start to see an effect on the number of cases and around 14–20 days to start to see an impact on deaths.

Spain has been one of the countries hardest hit by the coronavirus, both in some infections and deaths. The lack of homogeneity (and even the doubts on the integrity) in the accounting of the cases in the different countries, the limited knowledge about the nature of the Pandemic, and the different rates of evolution of this by country make comparison difficult. Instead of conclusions, only guesses can be made. In any case, it makes sense to inquire into the possible reasons why the spread has been so severe in Spain in light of the data acquaintances, and how to prepare for the future and of course, has not only uniquely suffered the Pandemic but also addresses the phase of reconstruction with specific significant weaknesses. Your dependence on tourism (one of the most affected sectors) joins the scarce technological base of its productive model, the fragility of the labor market, the entrenched inequality after the previous crisis, and at an earlier level of public indebtedness that conditions the stimulus plans. Furthermore, the internal political climate is characterized by polarization government-opposition, the scarce harmony between the territorial authorities, and a high distrust of citizens towards politicians.

In Europe during the two first weeks of March, several UEFA CHAMPIONS LEAGUE football matches took place even in Lombardia, where the virus infections started in Italy, Barcelona, and many other sites.

The schools and universities' lockdown in Spain produced a massive number of travels to other regions without any control. This had a dramatic effect on the spread of the virus in Spain's second residence houses.

This softer and more gradual response may have delayed the economic impact in the initial stage of the epidemic while promoting citizen cooperation through law. However, it has also led to an overload of health systems, evidenced by higher death rates than in China.

Furthermore, restrictions on the performance of diagnostic tests will lead to overestimated mortality rates and limit the value of available epidemiological information.

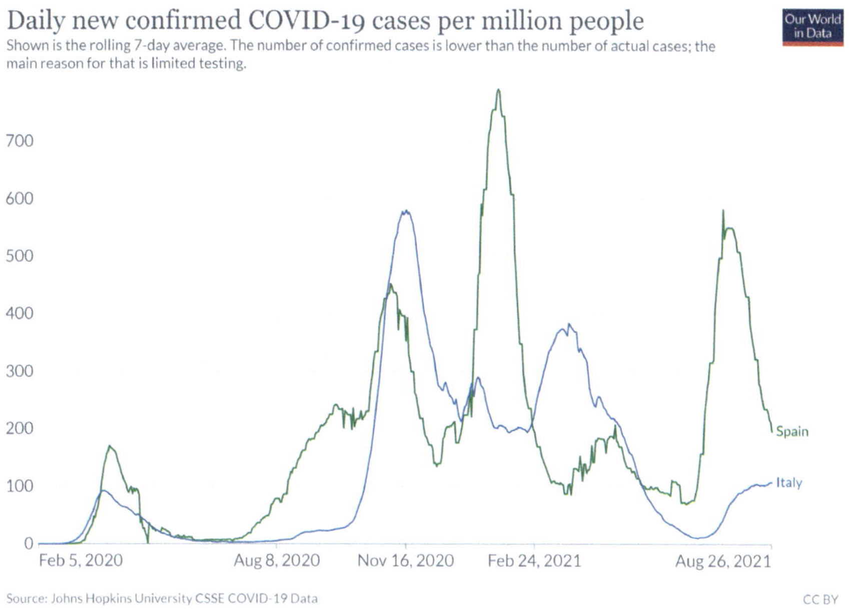

Daily new confirmed cases in Spain and Italy from February 5th, 2020/August 26th, 2021

The second wave arrived just after the summer without specific control, and the third and fourth peak just after Christmas celebrations. After the relaxation of the measures produced by the massive vaccination program in Italy and Spain, a fifth wave arrived caused by the delta variant.

> Measures in Italy and Spain have been somewhat mixed. Containment measures (contact and focus tracing) were insufficient, leading to an escalation of restrictive measures to mitigate the epidemic and "flatten the curve." Mass gatherings have been banned, schools have been closed, only certain types of work outside the home are allowed, and travel has been partially restricted. Unlike China, no tracking technology has been applied in Spain in the beginning. The curve may be beginning to soften in Italy, where a first closure of the northern provinces was announced on March 7th (more than a month after the first cases were reported on January 29th, although the virus is believed to have circulated since mid-January), followed shortly by a statewide shutdown. It is still too early to begin to see an impact from the blockade in Spain, which began on March 14th (a month and a half after the first cases were detected in the Canary Islands); It takes 5 to 7 days (the median incubation time) to start to see an effect on the number of cases and around 14 to 20 days to start to see an effect on deaths. (Chaccour, 2020)

But on the other hand, the global vaccination strategy of the EU has shown a significant advance in Spain and is leading the vaccination process in the EU.

EU mainly and also Spain specifically, for instance, is donating 7,500,000 vaccines to South America and Caribbean countries in need, through COVAX. The strategy, that has been proved as a good mechanism to distribute vaccines to the non-developed countries (Spain Government, 2021).

3.3 USA

In the USA, the President Trump administration created a problematic campaign of denialism about Pandemic

During the Pandemic's crucial early days and weeks, former President Donald Trump and other authority figures actively minimized the virus's threat.

There were inadequate tracing, isolating, and quarantines. However, the timeworn methods of combating an infectious disease—testing people who may be sick, tracing their contacts, and isolating or quarantining those who are positive or exposed—worked for COVID as well.

In addition to its test problems, the US did not do an adequate job of isolating those who were known or suspected to be infected (or had recently traveled to a high-risk area), tracing their contacts, or requiring quarantines for those who were exposed. Instead, China imposed extremely strict, city-wide quarantines.

Therefore, the state and the local response was what was expected with a federal structure like that of the United States, which does not, however, protect the disconcerting reaction of the national government. The White House's mistakes range from rhetoric to organization. After minimizing the risks, putting the "hunches" of the president before the information of the experts, the declaration of national emergency arrived late; the decision to ban travel from Europe initially excluding the UK and Ireland created uncertainty and much concern because it showed that the White House response was guided more by purely political issues than by evidence. The queues and chaos at US airports after the first air restrictions, complying with the recommendation of "social distance," showed a lack of coherence and a lot of improvisation of the Administration. It was ensured that everyone was given a diagnostic test when it was not possible; it was claimed that they were very close to the vaccine with months to go, and it was announced that Google had a website to do diagnostic tests when it was an idea not yet implemented. While some states have been asking the federal government to help with a shortage of gowns, gloves, and masks for days, governors were told to find a way to get ventilators and respirators because the federal government would not provide them. Governor Cuomo, for his part, implored federal support to build temporary hospitals in the face of the overflow of the state hospital network because only the national state has the capacity and resources to convert university residences and government facilities into hospitals: "The state can't do that." (Frost, 2020).

Other countries required those who may have been exposed to stay at a government-approved hotel or other facilities for a quarantine ranging from a few days to a couple of weeks. Such policies would likely have been harder to implement in the US, a nation that prides itself on personal freedoms. But not doing so came at the expense of keeping the virus in check.

Also, confusing mask guidance was at the beginning. Although face masks are now widely considered crucial in stopping transmission, US and global health authorities were slow to recommend them for public use.

Even after health experts concluded that masks were effective, Trump refused to set an example by wearing one in public.

Structural racism fueled health inequities. The Pandemic exposed and exacerbated deep-rooted racial and economic inequality in health. As a result, black and Hispanic individuals and other B.A.M.E. people caught COVID and died at disproportionately high rates.

Also, there was a decentralized response. The US government's structure meant that much of the pandemic response was left to state and local leaders. In the absence of a robust national strategy, states implemented a patchwork of largely uncoordinated policies that did not effectively suppress the spread of the virus. This caused sudden, massive spikes of infections in many local outbreaks, placing enormous strain on health care systems and leaving no region untouched by the disease.

Under normal circumstances, a crisis of these characteristics would have pushed the US to assume international leadership by mobilizing resources and bringing countries together to row in the same direction. That was the case after the tsunami in Southeast Asia, in the financial crisis of 2008, and after the Ebola outbreak in 2019. Now he no longer wants to be the conductor of an international community that he does not seem to want to be part of either. According to a German newspaper, President Trump tried to persuade a German company working on a coronavirus vaccine to move the research from Europe to the US. According to the newspaper, the intention was to have a vaccine just for "America eventually." It is not clear what happened, but given the current Administration's behavior, such behavior seems plausible. To this, we can add that in Italy, China, not the United States, sends medical equipment and assistance to a healthcare system at the limit. Or that the restrictions on flights from Europe were decided unilaterally without any prior consultation or coordination or even warning to Brussels and European partners. (Garcia Encina, 2020).

However, the warning signs that the US was not prepared to face a pandemic of these characteristics have been flashing for more than a decade, pointing mainly to its health system. Furthermore, throughout multiple administrations—not just the Trump Administration—US governments have not prioritized being prepared for a pandemic in advance. In general, the flow of funds that are used to mitigate or contain such a situation usually always comes after the crisis has broken out. In this sense, Donald Trump's emergency declaration follows the pattern of previous problems—SRAS, MERS, H1N1, Ebola, Zika—that released billions of federal funds until it was all over and forgotten. Although the US has had a national strategy for pandemic influenza for years, five strategy manuals do not usually include funding sustained over time to acquire a capacity to prepare and respond to an emergency. Although some, like Barack Obama, tried.

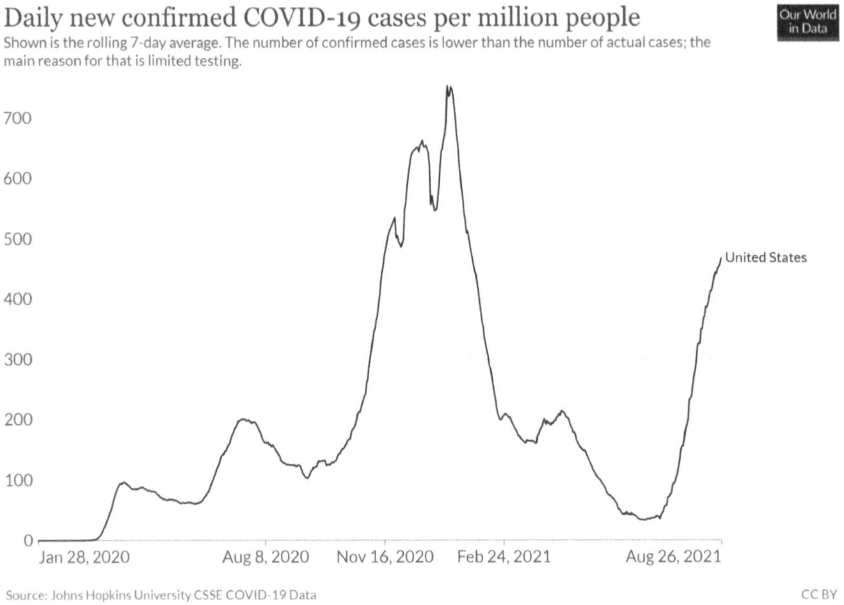

Daily new confirmed cases in the USA from February 5th, 2020/August 26th, 2021

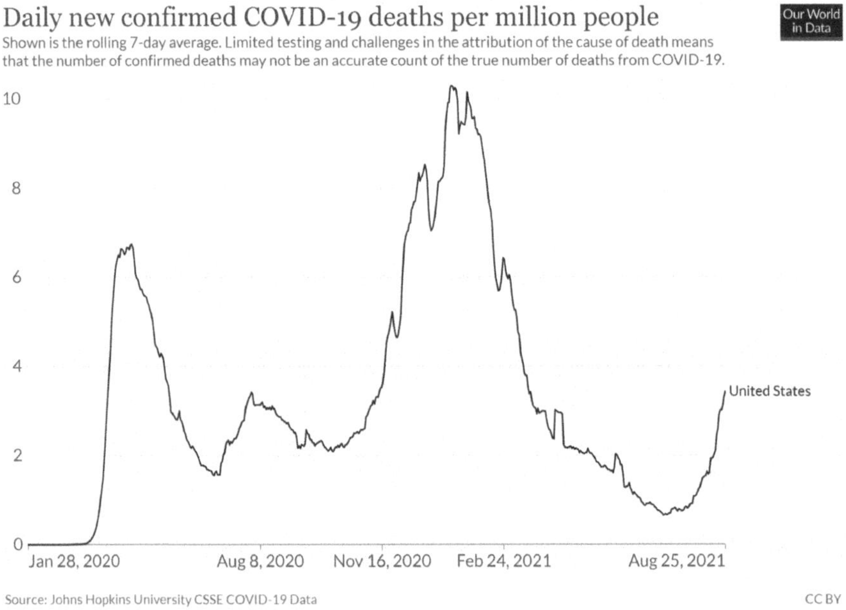

Deaths confirmed cases in the USA from February 5th, 2020/August 26th, 2021

4 Ethical Issues Regarding the Pandemic

The SARS-CoV-2 virus pandemic is causing a significant health crisis magnitude by the number of infected people who pose a risk to the rest of the population and by the high number of people who fall ill and require health care, very often hospitalized and critical, which requires extraordinary measures of various types that are projected to the entire population and those affected.

The imposition of the prevalence of general interests over individual interests in a pandemic situation, which may lead to the restriction or suspension of fundamentals rights of diverse nature, must not invade the actual content of these rights, being subject to the principles of equity, non-discrimination, solidarity, justice, and proportionality (Emanuel et al., 2020).

Health authorities oversee protecting the health of the population and respond to public health emergencies. Adequate health response to emergencies such as the current Pandemic, declared by the WHO of COVID-19, requires the most up-to-date information to obtain this information; the authority's health must carry out surveillance activities and act without delay based on the information received. In the context of the COVID-19 Pandemic, vigilance is necessary to reduce some of the uncertainty that characterizes it. Therefore, the health authorities must ensure that information is collected rigorously, that all relevant cases are notified and that the data are handled responsibly, always considering the benefit of the population (Rosembaum, 2020).

We all want a treatment that stops the coronavirus and a vaccine to protect ourselves. But biomedical research has its times and procedures for compelling reasons. Nothing less than to comply with the first two bioethics principles: the principle of non-maleficence (not doing evil) and the focus of beneficence (doing good).

These two principles remind us that we must evaluate the safety and efficacy of any treatment before authorizing it. The medicines, above all, should not cause more damage than we intend to solve. The benefits must outweigh the risks. They must also be effective for your goal. First of all, safe, then applicable. This benefit should also consider the opinion of the patient. This is related to the third principle of bioethics: the principle of autonomy (which requires respecting the freedom of decision of any patient about any intervention through so-called informed consent).

> Ethics are considered a basic aptitude in healthcare, and the capacity to handle ethical dilemmas in tough times calls for an adequate, responsible, and blame-free environment. While do-not-resuscitate (DNR) decisions are made in advance in certain medical situations, in particular in the setting of poor prognosis like in advanced oncology, the discussion of DNR in relation to acute medical conditions, the COVID-19 pandemic in this example, might impose ethical dilemmas to the patient and family, healthcare providers (HCPs) including physicians and nurses, and to the institution. (Sultan et al., 2021)

As in other surveillance cases, during the Pandemic, the authorities of public health may have to collect personal data or samples. Even though maybe It is not appropriate to obtain informed consent to collect these data, they must manage respectfully, protect people's privacy, maintain confidentiality as much as possible, and provide the information on transparently collecting data. Health authorities public also have an ethical duty to implement interventions that are already known that work. (Bavel et al., 2020).

But not all activities that comprise data collection in one way systematically constitute research with human beings. It is characterized by having the primary purpose of generating generalizable knowledge.

The Health authorities participate in different types of research. It is necessary to obtain ethical approval previously and in which participation is voluntary and preceded by an informed consent process. Health authorities (Council of Europe, 2021).

They also carry out activities whose main objective is the direct benefit of the population they serve, for example, improving their health or addressing health problems public. Even if those activities include systematic collection or analysis of personal data, as in the case of surveillance, they do not constitute investigation with human beings.

Therefore, they are not subject to the rules and regulations that govern research involving humans, such as prior investigation approval by an ethics committee. However, surveillance and other public health activities should be carried out ethically, for example, seeking to minimize risks to individuals and communities.

Ethical guidance and appropriate supervision, especially in the context of a health emergency public, should be used in this case of a pandemic.

But it is often difficult to distinguish between public health research and other public health initiatives and activities, particularly during a disaster emergency.

The Pandemic can pose thorny questions when it comes to the definition of priorities. We must anticipate scenarios in which health systems are saturated and may not give attention to all the people who need, for example, access to ventilators or intensive care beds to all affected patients. Health authorities have an ethical obligation to provide a public justification for the criteria used to define attention priorities. Transparency about the arguments for prioritization decisions improves public trust, increases acceptability, and promotes compliance with related recommendations.

In Spain, for instance, the last Spanish Constitutional Court decision on the July 2021 case 2054/2020 against the State of Alarm declared on March 2020 by the Spanish government due to the COVID 19 pandemic announced it was not legal the extraordinary restrictions on freedom of movement through the national territory imposed by article 7 (Sects. 1, 3 and 5) of RD 463/2020, even if they are oriented to the protection of constitutionally relevant values and interests, and comply with the measures recommended by the World Health Organization in its document "Update of the strategy against COVID-19" (April 14th, 2020), exceed the scope of the state of alarm recognized by the Constitution. The consequence of this decision was clear. The confinement during the State of Alarm is contrary to the Spanish Constitution, and all the sanctions imposed by the police were not valid (Tribunal Constitucional Spain, 2021).

The court has considered that some of the measures and restrictions that had been taken correspond more to a state of emergency than the state of alarm.

Similar questions are now known according to the deniers and the vaccines. In the case of France, several impositions of the French government, according to COVID 19, are being criticized by the people.

The COVID-19 Pandemic has brought to light or revealed real questions of an ethical nature, revealing the limits of knowledge, the limits of life, the limits of our health system, the limits of our society. Therefore, questioning these limits and these ethical questions can be interesting to advance our health system, our community.

Vaccination is also an issue with ethical implications that has been discussed before talking about the distribution of vaccines. 99% of people in poor countries have not been vaccinated (Jesus, 2021).

It is impossible to obtain global immunity without a global vaccination program, but luckily, initiatives though the WHO like COVAX, we hope may assure a significant progress to end this Pandemic after almost two years.

We have to address some ethical questions with regard more particularly to the function of care, the relationship to death, the connection to uncertainty, questions more related to containment measures than to the COVID itself, and finally, a more political questioning on what this epidemic reveal about the fragility of our societies, our economies. We arbitrarily set aside, given the constrained format of the article, specific fundamental questions such as questions relating to research ethics, questions relating to the notion of selection or sorting of patients, questions about the ethics of shortage management, and finally, issues related to tracing contact or sick persons (Aubry, 2020).

5 Conclusions

The concluding statements focus on the fundamental arguments demonstrated in the preceding text:

5.1. In conclusion, no single universal approach can effectively respond to today's rapidly evolving situation.

5.2. Each country must adopt the response according to the capacities of its health systems, its economic resources and infrastructure, and the degree of collective and individual responsibility and compliance with the recommendations issued by the authorities.

5.3. Effective control of COVID-19 requires governments and their constituencies to engage in mutually trusting relationships with a shared understanding of what is expected by both sets of actors.

5.4. The ability of government and public health leaders to gauge how the population perceives the effectiveness of government responses to COVID-19, both generally and on specific responsibilities, is essential for identifying potential obstacles to achieving disease control objectives.

5.5. National responses in a significant number of countries failed to get ahead of the Pandemic. Measures taken too late had all of the costs but none of the benefits of early containment, resulting in a negative feedback loop in which the economy was pitted against health.

5.6. The Vaccination campaign must be addressed through global cooperation to assure and gain global immunity and end the pandemic.

5.7. More decisive leadership and better coordination at national, regional, and international levels, including a more focused and independent WHO, a Pandemic Treaty, and a senior Global Health Threats Council is needed for a better resolution in the future.

5.8. Ethical issues in the Pandemic have not been fully respected, but it is not easy to find the limits to this situation.

5.9. Finally, we must consider that it has been a chain made up, at each point, of weak links. Pandemic preparedness was inconsistent and underfunded. The warning system was too slow and too tame. WHO did not have the necessary power. The answer has exacerbated inequalities with a world political leadership absent, which probably could have been avoided and persists today.

References

Arteaga, F. (2020). *La gestión de pandemias como el COVID-19 en España: ¿enfoque de salud o de seguridad? Real Instituto Elcano.* http://www.realinstitutoelcano.org/wps/wcm/connect/e2501bf2-2f67-47fc-a130-29624e129fa6/ARI42-2020-Arteaga-gestion-de-pandemias-COVID-19-en-Espana-enfoque-de-salud-o-de-seguridad.pdf?MOD=AJPERES&CACHEID=e2501bf2-2f67-47fc-a130-29624e129fa6

Aubry, R. (2020). *¿Quels enjeux de nature etique l'epidemie de COVID-19 a-t-elle soulevé?* https://www.ncbi.nlm.nih.gov/pmc/articles/PMC7315950/

Bonanno, G. A., Ho, S. M. Y., Chan, J. C. K., Kwong, R. S. Y., Cheung, C. K. Y., Wong, C. P. Y., & Wong, V. C. W. (2008). Psychological resilience and dysfunction among hospitalized survivors of the SARS epidemic in Hong Kong: A latent class approach. *Health Psychology, 27*(5), 659–667.

Centers for Disease Control and prevention (2020). *An approach for monitoring and evaluating community mitigation strategies for COVID-19.* Atlanta: Centers for Disease Control and Prevention. https://www.cdc.gov/coronavirus/2019-ncov/php/monitoring-evaluating-community-mitigation-strategies.html

Chaccour, C. (2020). *Cinco Respuestas de salud pública diferentes ante la epidemia.* https://www.isglobal.org/healthisglobal/-/custom-blog-portlet/covid-19-cinco-respuestas-de-salud-publica-diferentes-ante-la-epidemia/2877257/0

Council of Europe. (2021). *Consultative Committee of the convention for the protection of individuals with regard to automatic processing of personal data. Covid-19 vaccination, attestations, and data protection.* https://rm.coe.int/t-pd-bur-2021-6rev2-statement/1680a25713

Council of Foreign Relations (2021). A Guide to Global COVID-19 vaccine efforts. https://www.cfr.org/backgrounder/guide-global-covid-19-vaccine-efforts

Covid-19 Vaccines Global Access, COVAX. (2020). https://www.who.int/initiatives/act-accelerator/covax

De Jesus, M. (2021). *Global herd immunity remains out of reach because of inequitable vaccine distribution—99% of people in poor countries are unvaccinated.* The Conversation. https://theconversation.com/global-herd-immunity-remains-out-of-reach-because-of-inequitable-vaccine-distribution-99-of-people-in-poor-countries-are-unvaccinated-162040

Emanuel, E. J., Persad, G., Upshur, R., Thome, B., Parker, M., Glickman, A., Zhang, C., Boyle, C., Smith, M., & Philips, J. P. (2020). Fair Allocation of Scarce Medical Resources in the Time of Covid-19. *The New England Journal of Medicine*. https://doi.org/10.1056/NEJMsb2005114

European Centre for disease prevention and control (2021). Overview of the implementation of COVID-19 vaccination strategies and deployment plans in the EU/EEA. https://www.ecdc.europa.eu/en/publications-data/overview-implementation-covid-19-vaccination-strategies-and-deployment-plans

European Commission. (2020). EU Vaccines strategy. https://ec.europa.eu/commission/presscorner/detail/en/ip_20_1103

Frost, M. (2020). Cuomo warns coronavirus "wave" threatens hospitals, calls for federal help, *Brooklyn Daily Eagle,* 15/III/2020. "Cuomo warns coronavirus "wave" threatens hospitals, calls for federal help".

Gandhi, M., Yokoe, D. S., & Havlir, D. V. (2020). Asymptomatic transmission, the Achilles' heel of current strategies to control covid-19. *New England Journal of Medicine, 382*(22), 2158–2160.

García-Basteiro Al, H. (2020). *Legido-Quigley, on behalf of 20 signatories. Evaluation of the COVID-19 response in Spain: principles and requirements.* Lancet Public Health.

García-Basteiro, A. L., Legido-Quigley, H., & 20 signatories (2020). Evaluation of the COVID-19 response in Spain: Principles and requirements. The Lancet. *Public health, 5*(11), e575.

Garcia Encina, C. (2020). EEUU frente al COVID-19. Real Instituto Elcano. http://www.realinstitutoelcano.org/wps/portal/rielcano_es/contenido?WCM_GLOBAL_CONTEXT=/elcano/elcano_es/zonas_es/ari29-2020-garciaencina-eeuu-frente-al-covid-19

Ghebreyesus, T. A. (2021). Vaccine Nationalism harms everyone and protects no one. Foreign Policy. https://foreignpolicy.com/2021/02/02/vaccine-nationalism-harms-everyone-and-protects-no-one/

Hale, T., Angrist, N., Goldszmidt, R., Kira, B., Petherick, A., Phillips, T., Webster, S., Cameron-Blake, E., Hallas, L., Majumdar, S., & Tatlow, H. (2021). A global panel database of pandemic policies (Oxford COVID-19 Government Response Tracker). *Nature Human Behaviour*. https://doi.org/10.1038/s41562-021-01079-8

Johns Hopkins Coronavirus Resource Center. (2021). *COVID-19 Map* [Internet]. https://coronavirus.jhu.edu/map.html

Lazarus, J. V., Ratzan, S., Palayew, A., Billari, F.C., Binagwaho, A., & Kimball, S., et al. (2020). COVID-SCORE: A global survey to assess public perceptions of government responses to COVID-19 (COVID-SCORE-10). *Plos One, 15*(10), e0240011. https://doi.org/10.1371/journal.pone.0240011

Montani, J. (2020). *La pandemia es global, pero su respuesta es nacional*. https://agendapublica.es/la-pandemia-es-global-pero-su-respuesta-es-nacional/

Rosenbaum, L., (2020). Facing Covid-19 in Italy. Ethics, Logistics and Therapeutics on the Epidemic's Front Line. *The New England Journal of Medicine*. https://doi.org/10.1056/NEJMp2005492

Sibley, C. G., Greaves, L. M., Satherley, N., Wilson, M. S., Overall, N. C., Lee, C. H. J., Milojev, P., Bulbulia, J., Osborne, D., Milfont T. L., Houkamau, C. A., Duck, I.M., Vickers-Jones, R., & Barlow, F.K. (2020). Effects of the COVID-19 pandemic and nationwide lockdown on trust, attitudes toward government, and well-being. *American Psychologist, 75*(5), 618–630. https://doi.org/10.1037/amp0000662. Epub 2020 June 4th. PMID: 32496074.

Sultan, H., Mansour, R., Shamieh, O., Al Tabba, A., & Al Hussaini, M. (2021) *DNR and COVID-19: The Ethical Dilemma and Suggested Solutions.*https://doi.org/10.3389/fpubh.2021.560405

Spain Government. (2021) https://www.lamoncloa.gob.es/serviciosdeprensa/notasprensa/exteriores/Paginas/2021/260721-maec_sanidad_covax.aspx

The Independent Panel for Pandemic. (2020) Preparedness & Response, COVID 19 Make the last pandemic, May 2021, 4. Available at: https://theindependentpanel.org/wp-content/uploads/2021/05/COVID-19-Make-it-the-Last-Pandemic_final.pdf

Tribunal Constitucional Spain. (2021). *Caso 2054/2020.* Press Note. https://www.tribunalconstitucional.es/NotasDePrensaDocumentos/NP_2021_074/NOTA%20INFORMATIVA%20N%C2%BA%2074-2021.pdf

van Bavel, J. J., Baicker, K., Boggio, P. S., Capraro, V., Cichocka, A., Cikara, M., Crockett, M. J., Crum, A. J., Douglas, K. M., Druckman, J. N. Drury, J., Dube, O., Ellemers, N., Finkel, E. J., Fowler, J. H., Gelfand, M., Han, S., Haslam, S. A., Jetten, J., Kitayama, S., Mobbs, D., Napper, L. E., Packer, D. J., Pennycook, G., Peters, E., Petty, R. E., Rand, D. G., Reicher, S. D., Schnall, S., Shariff, A., Skitka, L. J., Smith, S. S., Sunstein, C. R., Tabri, N., Tucker, J. A., Van der Linden, S., Van Lange, P. A. M., Weeden, K. A., Wohl, M. J. A., Zaki, J., Zion, S. & Willer, R. (2020). Using social and behavioral science to support COVID-19 pandemic response. *Nature Human Behavior*

World Health Organization (2020, March). *Director-General's opening remarks at the media briefing on COVID19.* Available at: https://www.who.int/director-general/speeches/detail/who-director-general-s-opening-remarks-at-the-media-briefing-on-covid-19---11-march-2020

Axiological, Economic and Legal Challenges for the Functioning of the Energy Union in the Context of Energy Security of the European Union

Agnieszka Pach-Gurgul and Juliusz Piwowarski

Abstract Currently, more than half of energy used in European Union comes from an import. That makes the EU the biggest importer of primary energy in the world. Approximately 30% of every crucial imported energy carrier such as oil, natural gas, or coal comes from Russia. In the crisis, Russia might cut off these energy carriers supplies for the EU. In response to the 2009 Russian-Ukrainian gas dispute and its consequences for EU countries, on 25th February 2015 European Commission approved documents related to forming an Energy Union. That project aimed to provide energy security, especially for states dependent on energy supplies from only one source. Undoubtedly, energy security is a significant value in the package of values that includes EU security, and in this context, the national security of Poland (RP) and its energy component. However, member states of the EU are not commonly interested in that idea because of different national interests. Some of them maintain positive economic and political relations with Russia.

Keywords Energy · Security · Critical infrastructures · European union · Energy union

1 Introduction

Apart from competitiveness and environmental protection, energy security is one of the three priorities of the energy policy of the European Union. However, for many years, this had been solely a concept, not supported with sufficient actions. So far, the emergence of each new initiative concerning the energy security of the Union had lost to national approaches which are filled with protectionism. The confirmation of the above status can be found both in legal regulations, which provide the

A. Pach-Gurgul (✉) · J. Piwowarski
University of Public and Individual Security "Apeiron" Kraków or Apeiron University, Kraków, Poland

J. Piwowarski
e-mail: juliuszpiwowarski@apeiron.edu.pl

J. Cayón Peña (ed.), *Security and Defence: Ethical and Legal Challenges in the Face of Current Conflicts*, Advanced Sciences and Technologies for Security Applications,
https://doi.org/10.1007/978-3-030-95939-5_13

member states with the possibility of individual creation of their energy security and also in the existence of unilateral energy policies carried out by the EU member states, which often represent solely the interests of the strongest of them. Given the fact that the energy security makes up an element of a national security and the security culture, each member state, so far has been guaranteeing it independently for itself (for example by means of bilateral contracts for the supply of raw materials, new infrastructural projects, etc.) (Piwowarski & Trzciński, 2019). Given the increase of the energy prices,[i] the depletion of the countries' own resources and progressing dependence on the import of raw materials, as well as quite restrictive ecological requirements connected with economy decarbonization, the energy security, on account of its complexity and multifactorial character, started to be perceived as a common challenge for the entire EU. It was only after the Russian-Ukrainian gas crises in 2006 and 2009 and resulting threats showed that the need arose to work out a project which will guarantee the energy security to all European Union member states. This project is named the energy union. The core objective for its creation was the involvement of all the member states and the execution of the principle of speaking with "one voice" in all matters concerning energy. The project, in its initial version, assumed that further integration within the Energy Union will cause the evolution of the attitude of the member states away from a particular mindset, narrowed down to the national interest, towards the attitude oriented at the interest of the entire European Union as a whole and, thus, towards the common energy security. Therefore, the research objective of this paper is to find the answers to the following questions:

- What does the energy union consist in, what are its objectives and principles?
- Why do some member states support the concept of the energy union, whilst some do not?
- Will the energy union increase the energy security of the European Union?
- What are its chances of success and will it be efficient in its proposed shape?

These questions are significant with respect to the disparities between the legal regulations of specific EU countries and the objectives of the creation of the energy security and the practical actions of specific EU member states.

The core research objective of the chapter is to introduce the essence and concept behind the energy union and also to verify the following research hypothesis: the discrepancy of the national interests of the specific EU countries not only undermines the construction of the energy union, but also changes its character pushing it towards a climatic union.

2 The Evolution of the Concept of Energy Security Within the European Union—A Literature Overview

The need of safety, in Abraham Maslow's understanding, is one of the basic human needs. According to this psychologist, each subject must experience the sense of safety in order to pursue other needs. The pursuit of the need for security has great significance for satisfying other, higher needs (Mitchell & Moudgill, 1976; Frei, 2003).Security, according to its definition, was, for a long time, associated with the state guaranteeing the certitude of existence and survival. Currently, such a view is questioned as being too narrow or even too conservative. In the new understanding, security is supposed to mean not only the guarantee of an inviolable survival of a specific subject, but also a freedom for its development (Brooks, 2008). The concept of security may be considered mostly in three aspects: mental pillar, social and organizational pillar, material pillar. Acording to Piwowarski and Czajkowski the three aspects create *security culture*, that is a phenomenon which allows to achieve: "• effective control over possible threats of a certain subject that activates for him in a particular place and at particular time the state of optimal degree of threats, • Recovery of subject's security after it has been lost, • Optimization of levels of multi-sectoral development process of a security subject that heads towards the harmony of security sectors in the context of hierarchization of subject's objectives, • Effective stimulation in a social and individual scale awareness of existence of the ultimate need of a human, self-improvement and creation of trichotomous development, mental, social and material, due to the promoting of beliefs, motivations and attitudes that strengthen individual and collective action in favour of the potential of the autonomous defence (self-defence) of individual and group security subjects" (Piwowarski & Czajkowski, 2018). These three aspects also affect the creation of various types of security, whilst one of them, which is very important, is energy security which makes up a constituent of the generally understood security defined a national security of a country (Chester, 2010; Dyer & Trombetta, 2013). Guaranteeing energy security belongs to the strategic tasks of a state, making up a measure of its efficiency. In the literature on the subject, energy security is defined in diverse ways (Piwowarski, 2015). The multitude of definitions keeps evolving continually on account of the changing character of security as existence, state, process and phenomenon (Pach-Gurgul & Ulbrych, 2019). Many researchers agree that there is no consensus with regards to one, complete, universal definition of energy security which is, in fact, an equivocal, multifaceted and dynamic notion (Kruyt et al., 2009; Winzer, 2012; Narula & Reddy, 2015). The simplest definition describes energy security as the "accessibility of sufficient energy resources for an affordable price" (Yergin, 2006). This security consists of such elements as integration of the market (of crude oil, gas, electricity), resources diversification, security margin (e.g. as the stocks of energy raw materials). A more detailed definition talks about the availability of energy "at all times, in various forms, in sufficient amount and for a reasonable price and/or the price possible to be paid" (Månsson, 2014). Energy security can be defined also as "the availability of adequate energy at an affordable price, which is reliable both from the point of view of technological development and also from the perspective of human safety" (Wang et al., 2018; Sovacool, 1900, Augutis et al., 2012). The definition provided by European Commission in turn, points to the fact that energy security of the European Union is identified with the security of supplies and is understood as an unbroken access to energy at any moment, in sufficient amount and for a reasonable price and/or the price possible to be paid" (Commission of The European Communities, 2006; The Council of The European Union, 2004).

The issue of energy security in the EU had been for many years treated as an element of a more extensive policy, not very significant for the operation of the economy of a given country, considered to belong to the area of *low politics*, i.e. the group of issues defined as technical ones, absorbing the attention of officers outside the domain of the state strategic planning (Pach-Gurgul & Ulbrych, 2019). This was the outcome of a fact that after the end of the World War II, the core basis of the European energy production was hard coal and crude oils, whilst their availability led to the lack of their treatment in strategic categories. The notion of energy security started to operate within collective consciousness as late as in the 1970s, during the oil crises. However, this issue still did not gain the rank of the priority in the policy of the majority of the European Union member states (Pach-Gurgul & Ulbrych, 2019). It was only during the Russian-Ukrainian gas crises of 2006 and 2009 that the debate on the EU energy security was reassumed and gained the priority status, strategic for the operation of specific member states (Van de Graaf & Colgan, 2017). These crises also led to the identification of energy security with the security of the gas supplies, which in fact narrows down the problem as the security of supplies is merely one of the elements of energy security and does not reflect the complexity of the problem. With time, the EU countries understood that they were struggling with the issue of security in a more extensive and multi-faceted dimension, as it is pointed out by Jääskeläinen et al. (2018), Dieter Helm (2014), and also Christa Uusi-Rauva (2010). They all emphasize three dimension of the issue:

1. physical dimension—concerning the protection of the assets, infrastructure (gas lines, oil pipelines, transmission lines etc.), trading routes and providing a real alternative for the supplies executed so far and fast substitution when the need arises;
2. political dimension, as energy security is also based on foreign policy conducted by specific member states;
3. ecological dimension connected with the guarantee of energy security with simultaneous decrease of the harmful impact of the energy production on the natural environment, by means of using low emission raw materials and increasing energy efficiency and the share of renewable sources of energy in energy balances (Yergin, 2012).

The analysis of the energy security cannot neglect its threats and determinants. The literature on the subject mentions primarily physical and economic threats of the energy security as well as others:

- physical e.g. short-term or even permanent interruptions in the supply of energy from one source or one region;
- economic e.g. the dependence on the energy prices, costs of transport;
- other e.g. high requirements connected with the environmental protection which affect the production, use and supply of energy raw products.

The list of threats for the energy security, apart from physical, economic and environmental risks must also include political hazards, resulting from international circumstances and the loss of the state's influence on the infrastructure of the transmission and distribution of energy.

The European Union is now facing the necessity of reconciling apparently incompatible actions and objectives which determine the security of energy supplies. It turned out to be necessary to subject the markets to the processes of liberalization, and, at the same time, to pursue environmental objectives, including, in particular, the guarantee of the availability of energy with simultaneous counteracting climatic change.

3 The Change of the Perception of the Issue of Energy Security in the EU

Currently the EU is the largest importer of primary energy—more than a half of the energy consumed in the EU (61.0%) comes from import. The situation as very dangerous as the ratio of dependency on import of primary energy tends to increase, as in 2000, it was 56%. In 2019, the energy mix of in the EU, i.e. the scope of the available sources of energy consisted mostly of five diverse sources: oil products (including crude oil) (36%), natural gas (22%), renewable energy (15%), nuclear energy and solid fossil fuels (13% in both cases)—cf. Fig. 1.

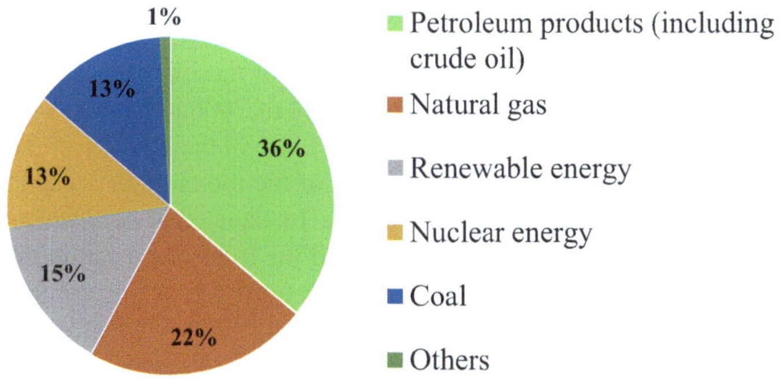

Fig. 1 European Union energy mix in 2019. *Source* own elaboration based on Eurostat: https://ec.europa.eu/eurostat/cache/infographs/energy/bloc-2c.html (access 21.08.2021)

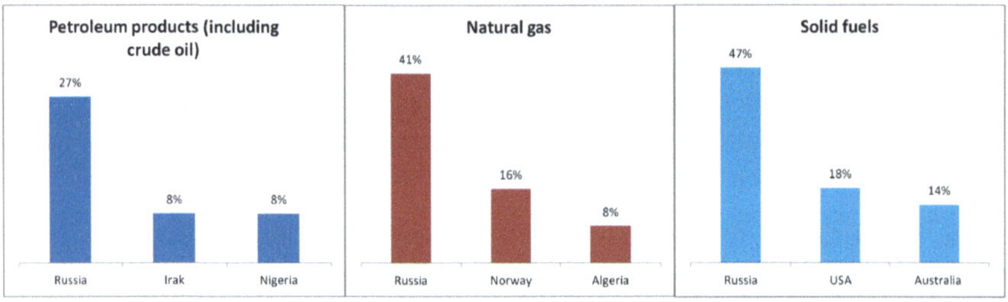

Fig. 2 The main suppliers (to the UE-28) of the key sources of energyin 2019. *Source* own elaboration based on Eurostat: https://ec.europa.eu/eurostat/cache/infographs/energy/bloc-2c.html (access 21.08.2021)

In the above mix, some of the raw products come from the import which is concentrated on relatively few external partners. In 2019, almost two thirds of the import of crude oil from outside the EU came from the following sources: Russia (27%), Iraq (9%), Nigeria and Saudi Arabia (8% each). A similar analysis shows that almost three quarters of the EU's import of the natural gas came from Russia (41%), Norway (16%), Algeria (8%) and Qatar (5%), and more than three quarters of solid fuels (mostly coal) came from Russia (47%), the United States of America (18%) and Australia (14%) (Eurostat, 2021). As it is shown in the graph below, the second dominating supplier of the raw products for the European Union is Russia, which, for some member states, supplies even as much as 100% energy raw products (Pach-Gurgul, 2017) cf. Fig. 2.

It is estimated that the dependence on the import of raw products will be progressing in the EU and therefore—the priority of the European policy with regards to the energy security will be the creation of an homogeneous energy market and solidarity of the member states in relations with the suppliers of the energy raw products. So far, the actions of the European Union in this respect have been insufficient. The EU countries have not worked out any coherent policy of the energy security with their actions in this regard being mainly individual, in accordance with the principle of subsidiarity.

The significant events which drew the attention of the EU member states to the problem of the energy security were the gas crises of 2006 and 2009. In particular, the gas crisis in winter 2009 illustrated the importance of the security of the supplies of the energy raw products to the EU, and also of the too large dependence on one supplier, i.e. Russia. The European Union, for 14 days, was deprived of the gas supplies at the level of 300 million m^3 daily and the losses incurred in this respect for the EU countries, and, in particular for the South-Eastern Europe amounted to approx. 1.6 billion EUR. The countries of the Central Europe were affected the most severely by the interruptions in the gas supply—only the Czech Republic imported some amount of gas via system of German gas lines system, which compensated for the deficiencies in the Russian supplies. The remaining countries could only rely on their accumulated strategic reserves. The situation was the most harmful for Slovakia (Müller-Kraenner, 2007)—for the first time in its history it was completely cut off from the external supplies of gas. Poland, in turn, did not receive the 80% of the planned gas supplies, whilst Hungary—60% (Pach-Gurgul, 2018). The crisis revealed also a varied level of the dependence of the European Union member states from the gas supplies from Russia: in the case of the Baltic states and Finland, this ratio was 100%, whereas in the case of France—only 19%, and for Sweden and Denmark there is no such dependence at all.[ii] This import differentiation and also other national interests with the Russian Federation (preferential contracts for energy raw materials, the construction of the Nord Stream 2 by Germany) make it difficult for the European Union to take a unified stance towards Russia, and, this distort the perception of the energy security within the entire Union. For the countries which import Russian gas for preferential prices (Great Britain, the Netherlands, Germany) this is a good solution which strengthens their energy security. For the Baltic states and Central-Eastern Europe, which are strongly dependent on the supplies of the Russian gas, and additionally, which purchase it for high prices, the import of this raw product from Russia is perceived as a threat for their energy security (Pach-Gurgul, 2018). It may be therefore, stated that within the European Union there is a clear distinction between the group dependent on one source of the supplies of the natural gas and also the other group—the countries in which the diversification of suppliers is more balanced. This division results in the fact that these countries have been perceiving the problem of energy security differently, which renders working out a common concept of energy security extremely difficult.

It seems that the surprise caused by Russia's cutting off the gas for Ukraine, felt by some countries of the European Union created a strong impulse to create the mechanism increasing the energy security for the member states and providing for other issues connected with its guarantee. There was an expectation that the member states would consider the interest of Europe's common energy security and pursue its execution in their foreign relations. This solution for the European Union was supposed to be the project of the energy union, which was intended to be a security buffer for the EU countries and some reliable remedy for the increasing divisions between the member states.

4 The Concept of the Energy Union of the EU Member States

The intention to increase the collaboration with regards to the energy security within the European Union is, first of all, identical for the countries of the Central and Eastern Europe and results from the belief that the increase of the collaboration will solidify their position within the EU and will make their voice resonate louder on the EU forum. This standpoint is determined, among others, by the infrastructure which they have at their disposal (oriented at the off-take of the natural gas from the East) and by the higher gas prices offered by Russia in comparison to the Western EU countries (Pach-Gurgul, 2017). The countries of the Central and Eastern Europe emphasize that the contemporary challenges and threats of the energy security force them to the collaboration which exceeds the bilateral relations. As a result of the economic costs which were the outcome of the Russian-Ukrainian gas crises in 2006 and 2009, these countries started to promote the concept of the energy union. Its initial objective was to provide the mechanism of energy security mainly for the countries dependent on the supplies of energy raw products (such as gas) from one side only, which weakened their negotiating position in case of negotiating contracts for the supplies of natural gas (Pach-Gurgul, 2018).

One of such initiatives was the proposal made in 2009 by Donald Tusk—the idea of the energy union as the element integrating the concept of the energy security of the European Union. According to this proposal, the energy union was supposed to be based on six pillars, comprising:

1. joint purchase of raw products—the foundation of one European institution which would purchase gas for all the 28 EU member states, which would allow to decrease the disproportion in the gas purchase prices in the EU;
2. mechanisms of energy solidarity—the introduction of the principle that if one or more EU countries are facing the interruption of the gas supply, other countries will provide the assistance;
3. development of the energy infrastructure, financing by the EU up to 75% necessary investments (gas storage tanks, pipelines) in the countries which are the most dependent on the supply of the Russian gas;
4. diversification of the energy carriers, the introduction of the necessity to use the domestic energy carriers, mostly coal and also shale gas;
5. strengthening the EU position in outside relations with regards to the energy policy, signing agreements for the purchase of gas (in a liquid form) from the exporters outside the European Union—USA, Algeria,
6. strengthening the EU position in outside relations by means of solidifying, by the EU, the Energy Community created 2005 with its Eastern partners, in order to expand the European gas market towards the east (Leal-Arcas, 2016). The energy union, in its original version was an attempt to centralize the issues related to natural gas and to shift the point of gravity in this area from the EU member states onto the EU institutions.

On 25th February 2015, the European Commission officially adopted the package concerning the creation of the energy union (European Commission, 2015). This document postulates the creation of the energy union based on climatic policy and concerns, first of all, the electricity market and, in some issues, also the natural gas. The package comprises three communications:

1. Framework strategy for the energy union—containing the objective of the energy union and specific methods of its creation;
2. The EU vision of the new global climatic agreement—this agreement was supposed to be settled in December 2015 in Paris;
3. The methods of reaching, by 2020, the target level of 10% electricity in interconnections.

The framework strategy for stable energy union (European Commission, 2015), defines three long-term goals for the EU energy policy: the security of supplies, sustainable supplies and competitiveness. The priorities of the energy union, which the Polish government promoted, were only partly reflected in the draft proposed by the European Commission, finally adopted by the member states in 2015. The proposal of the Polish government focused mainly on the union in the gas and crude oil sector and also on complete use of the domestic energy carriers, such as coal and shale gas. The adopted package, however, focuses, first of all, on the issues concerning electricity and also on the development of the renewable and low-carbon energy sources. Its core objective is made up of five, closely connected issues:

1. Energy security, solidarity, confidence

The objective is the make EU resistant to energy crises and to decrease its dependence on specific fuels, suppliers and routes of supply.

2. Internal market of energy

A new impulse is needed to complete the works connected with the creation of the internal market of energy—better interconnections, complete implementation and the execution of the current energy regulations.

3. Energy efficiency as a method of decreasing the energy demand.
4. Decarbonization of economy.

According to the strategy the Union was supposed to become the world leader of renewable energy and a global center of the works on the new, technically advanced and competitive sources of renewable energy.

5. Scientific research, innovations and competitiveness

The EU should be the leader in the technologies of energy networks and smart houses, in ecological transport, in clean fossil fuels and nuclear energy. For each of these areas, a directive was drafted and published, pointing to the legal possibilities of the realization of their objectives (Bellantuono & Huhta, 2019). The strategy concerning energy union, comprised a very limited form of the concept of joint purchase of gas, which was advocated by Poland. The document adopted by the European Commission mentions only that voluntary joint purchases of gas by groups of enterprises will be taken into consideration. The precondition for such purchase must be dependence on one supplier and, also the occurrence of a crisis in supplies. The European Commission emphasizes that such actions must comply with the EU anti-cartel measures and also with the provisions of the World Trade Organization (Pach-Gurgul, 2018).

The further part of the energy union package, comprised in the document, *Paris Protocol—counteracting the climate change in the world after 2020* (European Commission, 2015a), presented the EU vision of a new global climatic agreement, which was planned to be adopted and finally adopted in December 2015 in Paris. The document defined the objective providing for the decrease of greenhouse emissions by 40% by 2030. Also, the governments agreed to stop the world mean temperature on the level much below 2 °C, in relation to the level from the pre-industrial period and to try not to make it more than 1.5 °C (Pach-Gurgul, 2018). The last part of the energy union package proposed the methods to reach, by 2020, the target level of 10% electricity in interconnections. Therefore, it is extremely significant to:

- To improve the situation in 12 member states, in which the interconnections do not reach 10% (Ireland, Italy, Romania, Portugal, Estonia, Latvia, Lithuania, Great Britain,[iii] Spain Poland, Cyprus and Malta);
- To execute the projects planned within the TEN-E regulation and Connecting Europe Facility (CEF), which help to increase the interconnections;
- To provide for the possibility of financial support for the projects concerning interconnections and also regional support in energy production.

According to the European Commission, a correctly interconnected European energy network could bring the consumers the savings up to 40 billion EUR per year.

5 The execution of the Project of Energy Union—Main Challenges

In many member states expecting definite solutions in the matter of energy security, there are doubts concerning the sense of creating energy union. The departure from the initial form of the project showed again that there are disparities between the energy-related interests of the countries of Central and Eastern Europe, which are strongly dependent on the gas supplies from Russia and also France, Germany or the ecological Denmark. Therefore, Maroš Šefčovič, the Commissioner for the Energy Union, organized the Energy Union Tour, which was a series of meetings in all the EU member states, taking place at the turn of 2015 and 2016 (Pach-Gurgul, 2018). During the visits in the member states, he tried to convince them that the energy union would have a significant influence on the energy security of the EU, by means of the following (European Commission, 2017):

- Diversification of the direction of the supply of gas and other raw materials;
- Development of gas and electricity connections;
- Solidification of the solidarity mechanism during the crisis between the member states;
- Support for the investments in the energy sector;
- Support for the Projects of Common Interest.

Although the very concept of the energy union, in its original form seemed interesting, as some energy security buffer for the Union economy, the conclusions concerning its effect on energy security, which were formulated after the end of the meetings organized by Maroš Šefčovič in all the member states, seem to be very general. The critics of the project observe that perhaps it should be named rather a climatic union and then it would perfectly fit into a larger EU project—the European Green Deal. After a series of meetings, it was agreed that each year there would be a report published concerning the progress and challenges in the execution of the objectives of the energy union. So far, 5 reports have been published, out of which the last one was the most eagerly awaited. It comprises the account of the first months of the coronavirus pandemic and its effect on the execution of the energy union concept. The European Union, likewise the entire world, were placed in completely new conditions of the pandemic of coronavirus, in which, a concern arose that the execution of the objectives of the energy union may be given a lower priority, being superseded by such challenges as unemployment, the breakdown of the healthcare system and broken supply chains. However, in the report from 2020, the European Commission stressed that the member states were able to meet these objectives, and, in majority, they made satisfactory progress in their execution. In the opinion of the Commission, the report shows that the energy sector may contribute to the restoration of the Union economy after the economic crisis caused by the coronavirus pandemic. So far, the energy union proves to be strongly resistant to all the shocks resulting from the pandemic, whilst the employment in the energy sector is not disturbed. The European Commission points out that the report contained the analysis of five different dimensions of the energy union—decarbonization, including the renewable sources of energy, energy efficiency, energy security, internal energy market and scientific research, innovations and competitiveness. The document contains also the instructions for the fast implementation of national plans for energy and climate and for the recovery of the Union economy with the use of large scale investments and reform in the energy production sector to be carried out in the difficult times of the pandemic. The report takes into consideration the first months of the pandemic and thus, a question arises whether 2021 will also allow for such optimistic conclusions in the execution of the energy union project and how these conclusions will be affected by the execution of the plan of the EU economy renewal. The long-term EU budget in connection with the NextGenerationEU (NGEU), a temporary mechanism triggering the economy renewal is the largest package of the financial resources designated for development. The total amount of 2.018 trillion EUR in the current prices will be used for the reconstruction of Europe after the COVID-19 pandemic. Thanks to this amount, the New Europe will be more environment-friendly, more digital and more resistant to crises (European Commission, 2021).

6 Conclusions

After the gas crisis in 2009 and the disclosure of the weakness of the solidarity mechanism of the European Union, a need arose to strengthen the energy security. This challenge was supposed to be met by the creation of the energy union. The original form of the project provided for the creation of a European institution which would purchase gas for all the 28 EU member states (after the Brexit for all the 27 countries), which would thus allow for the increase of the level of supplies and reduction of the disparities in the purchase prices within the EU. The project's objective was also to use the domestic energy carriers, mostly coal and also shale gas. Today, it is already known that the suggestion of joint purchase was rejected and the concept of the return to coal contradicts the idea of economy decarbonization. The project of the energy union, in its final form, concentrates on energy security, making the EU resistant to the external energy crises and the decrease of its dependence of specific fuels, suppliers and routes of supply. The solutions comprise mostly: the development of the internal energy market, increase of energy efficiency, decarbonization of economy scientific research, innovations and competitiveness. Unfortunately, the project does not comprise the complex strategy of energy security which would be satisfactory for all the EU countries, which is the outcome of two diverse approaches to the issue represented by the countries of the western and central-eastern Europe. The countries and companies from the western part of the EU believe that the gas market is liquid enough for them and it should be influenced by the business and purely economic factors. However, the countries of the Central and Eastern Europe which import gas, first of all, from Russia, have too limited infrastructure and diversification options with regards to the gas supplies from other sources. During Ukrainian-Russian crises, these states were often in very difficult situations (they were cut off from the gas supply), which led to significant burdens for their economies. Therefore, the objectives of the energy union do not encounter a similar level of interest of all the EU member states. This is also connected with the national interests of these states, which, for many years, have had good relations with the Russian Federation not only with regards to the energy raw products. The member states with developed energy infrastructure and sufficient stock of energy raw materials are less willing to change the mechanisms of collaboration. It seems that each member state still has its own vision of providing the energy security and this fact may weaken the practice of the execution of the energy union, in particular that, in accordance with article 194 of the Lisbon Treaty, energy production, and also the issues connected with the energy security rest with the member states. This may result in the fact that these countries will still pursue their particular interests connected with providing energy security, without thinking along the terms of the entire European Union. As a consequence, the energy union may not play such as important role in guaranteeing the energy security.

This complex situation is the outcome of the co-existence of many factors, including the treatment of energy production as the sphere resting solely with specific countries. In connection with the climatic policy, conducted simultaneously on the EU level, a regulatory chaos was made: the community climate objectives create framework for the national objectives of the energy policies, which automatically gives the priority to the first ones, at the cost of the priority of the energy security of the EU. Moreover, currently, in the period of the coronavirus epidemic, it may turn out that the problem of energy security and the execution of the objectives of the energy union will be sidelined, superseded by such problems as unemployment, the bankruptcy of many companies, health policy in the EU countries and digitalization. Fortunately, however, the initial reports do not confirm these concerns, as the energy security, access to energy and continuity of the supplies of raw products and energy with affordable prices allow rather for solving the above problems and may guarantee the development and renewal of the EU economy.

Notes

(i) With regards to the threats related to the COVID-19 disease caused by the coronavirus, the majority of the world countries affected by the pandemic, initiated isolation actions which decreased their production and service activity causing the decline of the demand for crude oils, which, as a result led to the decrease of the crude oil prices (in April 2020 the price reached the lowest level in more than 30 years).

(ii) Eurostat, 2018, *Complete Energy balances*, http://appsso.eurostat.ec.europa.eu/nui/show.do?dataset=nrg_110a&lang=en (access: 3.04.2017).

(iii) In the analyzed period, Great Britain was still in the European Union.

References

Augutis J., Krikstolaitis R., Martisauskas L., & Peciulyte S. (2012). Energy security level assessment technology, *Applied Energy, 97*, 143–149. https://doi.org/10.1016/j.apenergy.2011.11.032.

Bellantuono, G., & Huhta, K. (2019). *The Energy Union in the Next Decade*, OGEL 3. https://www.ogel.org/article.asp?key=3824. Access 21 Aug 2021.

Brooks, D. J. (2008). What is security: Definition through knowledge categorization, July 2010. *Security Journal, 23*(3), 225–239. https://doi.org/10.1057/sj.2008.18

Chester, L. (2010). Conceptualising energy security and making explicit its polysemic nature. *Energy Policy, 38*(2), 887–895. https://doi.org/10.1016/j.enpol.2009.10.039

Dyer, H., & Trombetta M. J. (2013). The concept of energy security: broadening, deepening transforming. In: H. Dyer & M.J. Trombetta (Eds.), *International handbook of energy security, Northampton* 2013, pp. 3–18.

European Commission. (2015a). Energy Union Package. Communication from the commission to the European parliament, the council, the European economic and social committee, the committee of the regions and the European investment bank. A framework strategy for a resilient energy union with a forward-looking climate change policy, COM(2015) 80 final, Brussels 25.02.2015.

European Commission. (2015). *Energy Union Package. Communication From the Commission to the European Parliament and the Council. The Paris Protocol—A blueprint for tackling global climate change beyond 2020*, COM(2015) 81 final, Brussels 25.02.2015.

European Commission. (2017b). *Benefits of the Energy Union—country factsheets*. http://ec.europa.eu/priorities/publications/benefits-energy-union-country-factsheets_en (access: 5.04.2021).

European Commission. (2021). *The EU's 2021–2027 long-term budget & Next Generation EU. Facts and figures*. Directorate-General for Budget, published : 29 Apr 2021.

Eurostat. (2021). *Complete Energy balances*, http://appsso.eurostat.ec.europa.eu/nui/show.do?dataset=nrg_110a&lang=en. Access: 13 Aug 2021.

Frei, C. W. (2004). The Kyoto protocol—a victim of supply security? or: if Maslow were in energy politics. *Energy Policy, 32*(11), 1253–1256.

Helm, D. (2013). The European framework for energy and climate policies. *Energy Policy, 64*, 29–35. https://doi.org/10.1016/j.enpol.2013.05.063.

Jääskeläinen, J. J., Höysniemi, S. H., Syri, S., & Tynkkynen, V. P. (2018). Finland's dependence on Russian energy—Mutually beneficial trade relations or an energy security threat. *Sustainability, 10*(10), 1–25. https://doi.org/10.3390/su10103445.

Kruyt, B., Vuuren, D. P. van Vries, H. J. M., & de Groenenberg, H. (2009). Indicators for energy security. *Energy Policy, 37*(6), 2166–2181. https://doi.org/10.1016/j.enpol.2009.02.006.

Leal-Arcas, R. (2016). *The European energy union: The quest for secure, affordable and sustainable energy* (European Energy Studies, vol. 8).

Månsson, A. (2014). Energy, conflict and war: towards a conceptual framework, *Energy Research & Social Science, 4*, 106–116. https://doi.org/10.1016/j.erss.2014.10.004.

Mitchell, V. F., Moudgill, P. (1976). Measurement of maslow's need hierarchy. *Organizational Behavior and Human Performance, 16*(2), 334–349. https://doi.org/10.1016/0030-5073(76)900 20-9.

Müller-Kraenner, S. (2007). *Energy Security. Re-Measuring the World*. London

Narula, K., & Reddy, B. S. (2015). Three blind men and an elephant: The case of energy indices to measure energy security and energy sustainability. *Energy, 80*, 148–158.

Pach-Gurgul, A., & Ulbrych, M. (2019). Progress of V4 Countries towards the EU's climate and energy targets in the context of the energy supply security improvement. *Entrepreneurial Business and Economics Review, 7*(2), 175–197. https://doi.org/10.15678/EBER.2019.0702.

Pach-Gurgul A. (2017). The energy situation of visegrad group countries in context of energy union. In *The Proceedings of the 17th International Joint Conference Central and Eastern Europe in the Changing Business Environment* (pp. 180–192). Bratislava, ISSN 2453-6113.

Pach-Gurgul, A. (2018). The energy union tour—Success or failure? *Political Science Rewiev, 1*, 103–118. https://doi.org/10.14746/pp.2018.23.1.7

Piwowarski, J., & Czajkowski, W. (2018). Cross-Cultural dialogue as a conflict management strategy in: cross-cultural dialogue as a conflict management strategy. In J. Martin Ramirez & G. Abad (Eds.), Springer International.

Piwowarski, J., & Trzciński, Ł. (2019). Anthropological aspects of security culture analyses within the context of selected elements of threat and security typology. *Security Dimensions, 32*, 108–126. https://doi.org/10.5604/01.3001.0014.1149.

Piwowarski, J. (2015). *Fenomen bezpieczeństwa*, WSBPI "Apeiron", Kraków.

Sovacool, B. K., Mukherjee, I., Drupady, I. M., & D'Agostino, A. L. (2011). Evaluating energy security performance from 1990 to 2010 for eighteen countries. *Energy, 36*,(10), 5846–5853. https://doi.org/10.1016/j.energy.2011.08.040.

The Council of the European Union. (2004). Council Directive 2004/67/EC of 26 April 2004 concerning measures to safeguard security of natural gas supply. *Official Journal of the European Union*, L 127/92, 29.4.2004.

Uusi-Rauva, C. (2010). The EU energy and climate package: a showcase for European environmental leadership? *Environmental Policy and Governance, 20*(2), 73–88. https://doi.org/10.1002/eet.535

Van de Graaf, T., & Colgan J. (2017). Russian gas games or well-oiled diplomacy? Energy security and the 2014 Ukraine crisis. *Energy Research & Social Science, 24*, 59–64.

Wang, B., Wang Q., Wei, Y. -M., & Li, Z. -P. (2018). Role of renewable energy in China's energy security and climate change mitigation: An index decomposition analysis. *Renewable and Sustainable Energy Reviews, 90*, 187–194, https://doi.org/10.1016/j.rser.2018.03.012.

Winzer, C. (2012). Conceptualizing energy security. *Energy Policy, 46*, 36–48. https://doi.org/10.1016/j.enpol.2012.02.067.

Yergin, D. (2012). *The quest: Energy, security, and the remaking of the modern world*.

Yergin D. (2006). Ensuring energy security. *Foreign Affairs, 85*(2), 69–82.

Freedom of Speech in Times of Crisis: The Case of Spain During the COVID-19 Pandemic

Sonia Boulos

Abstract The aim of this chapter is to analyze the impact of the COVID-19 pandemic on freedom of speech in Spain. It inquires whether the Spanish government and regional governments alike have met their obligation under international law to respect and ensure freedom of speech during the pandemic. The chapter identifies structural problems in the legal protection of freedom of speech in Spain that existed prior to the pandemic and demonstrates how they opened the door for the authorities to impose excessive limits of freedom of speech.

Keywords COVID-19 · Freedom of speech · Freedom of information · Disinformation · State of emergency

1 Introduction

The COVID-19 crisis has triggered intense debates, both locally and at the international level, on crucial issues ranging from healthcare policies, economic policies, access to education and so on. These debates were and remain a matter of utmost public interest and the media has a central role in facilitating them (Noorlander, 2020). However, lockdowns and restriction on the freedom of movement, as well as measures adopted to counter disinformation in relation to the pandemic, had a negative impact on the ability of the media and the public at large to be well informed on these crucial issues. Additionally, many States used the COVID-19 pandemic as a pretext to crackdown on freedom of speech with the aim of silencing public criticism of governments. According to Human Rights Watch (HRW), since January 2020, at least eighty-three governments have used the COVID-19 pandemic as an opportunity to silence critics and adopt new repressive laws criminalizing speech (HRW, 2021). Among the violations identified by HRW (2021) are: physical attacks on journalists, bloggers, and protesters by the police and other security forces for criticizing the State's response to the pandemic; utilizing Covid-19 and pre-Covid-19

S. Boulos (✉)
Antonio de Nebrija University, Madrid, Spain
e-mail: sboulos@nebrija.es

J. Cayón Peña (ed.), *Security and Defence: Ethical and Legal Challenges in the Face of Current Conflicts*, Advanced Sciences and Technologies for Security Applications,
https://doi.org/10.1007/978-3-030-95939-5_14

laws to arbitrarily arrest, detain, prosecute, and fine or imprison those who criticize the government's response to the pandemic; arbitrarily banning or breaking up protests against the government's responses to the pandemic; censoring media and social media coverage of the pandemic; enacting vague laws and measures criminalizing the spread of alleged misinformation or other coverage of the pandemic; and suspending or restricting the right to request and receive public health information, or granting press accreditation for Covid-19-related press briefings to pro-state media outlets.

While the Global Expression Report of 2021, which tracks freedom of expression across the world in 161 countries, suggests that Europe is the home to some of the world's most established democracies and highest-scoring countries on freedom speech, it demonstrates that in times of global health emergency, even States with strong record on freedom failed to confront the pandemic while fully respecting international human rights standards on free speech (Article 19, 2021). According to this report, Spain occupies the 19th position of high-scoring states on freedom of speech. Still, the report identifies violations of international human rights standards in relation to freedom of speech and freedom of information that occurred in Spain in 2020 (Article 19, 2021).

The aim of this chapter is to analyze the impact of the COVID-19 pandemic on freedom of speech in Spain. It delves on the question whether the Spanish authorities have met their obligation under international law to respect and ensure freedom of speech during the pandemic and attempts to identify gaps and deficiencies in the legal protection of freedom of speech in Spain that impacted negatively on this basic right during the pandemic.

The chapter is divided into three main parts. The first part discusses the protection of free speech in international law, including the limitations that could be imposed on free speech. The second part of the article identifies violations and threats to freedom of speech in Spain recorded during the pandemic. The Third part of the chapter identifies structural legal deficits that have contributed to limiting certain aspects of the right to free speech in the context of the pandemic. The chapter concludes that the limitations imposed during the pandemic run the risk of creating an environment that is detrimental to the exercise of political speech, even after the pandemic is over.

2 Freedom of Speech in International Law

As early as 1946 the UN General Assembly declared at its first session, that freedom of expression is "is a fundamental human right and is the touchstone of all the freedoms to which the United Nations is consecrated" (UN, 1946, preamble).

Freedom of speech is one of the central human rights protected by international human rights instruments. Article 19 of the International Covenant on Civil and Political Rights (ICCPR) (UN, 1966) states:

1. Everyone shall have the right to hold opinions without interference.
2. Everyone shall have the right to freedom of expression; this right shall include freedom to seek, receive and impart information and ideas of all kinds, regardless of frontiers, either orally, in writing or in print, in the form of art, or through any other media of his choice.

Freedom of speech is also protected by Article 10 of the European Convention of Human Rights (ECHR), adopted by the Council of Europe (COE):

1. Everyone has the right to freedom of expression. This right shall include freedom to hold opinions and to receive and impart information and ideas without interference by public authority and regardless of frontiers. This Article shall not prevent States from requiring the licensing of broadcasting, television or cinema enterprises (COE, 1950).

The Human Rights Committee (HRC), which monitors the implementation of the ICCPR by member states emphasizes that "[f]reedom of opinion and freedom of expression are indispensable conditions for the full development of the person. They are essential for any society. They constitute the foundation stone for every free and democratic society" (CCPR, 2011a: para. 2). It further emphasizes that freedom of speech is "a necessary condition for the realization of the principles of transparency and accountability that are, in turn, essential for the promotion and protection of human rights" (CCPR, 2011a: para. 3).

The European Court of Human Rights (ECtHR) (1976: para 49) has long stated that freedom of expression "constitutes one of the essential foundations of [a democratic] society, one of the basic conditions for its progress and for the development of every man".

Also, the Inter-American Court of Human Rights (IACtHR) (1985: para 70) delved on the democratic significance of freedom of speech in one of its advisory opinions:

> Freedom of expression is a cornerstone upon which the very existence of a democratic society rests. It is indispensable for the formation of public opinion. It is also a *conditio sine qua non* for the development of political parties, trade union, scientific and cultural societies and, in general, those who wish to influence the public. It represents, in short, the means that enable the community, when exercising its opinions, to be sufficiently informed. Consequently, it can be said that a society that is not well informed is not a society that is truly free.

Freedom of speech has been described as the cornerstone of all human rights. It is essential for the enjoyment of a wide spectrum of human rights, including right to assembly and political participation, and social, cultural and economic rights (McGoldrick, 2017; O'Flaherty, 2012). As O'Flaherty (2012: 630–631) points out:

> The high importance accorded to freedom of expression in international law and related discourse is not just a matter of philosophy or ideology. As a matter of empirical observation, it can be seen that free expression is essential to the good working of the entire human rights system.

Despite its centrality, or perhaps because of its centrality, the right to free speech is one of the most violated human rights on a global scale (O'Flaherty, 2012).

Freedom of speech is not an absolute right. Article 19(3) of the ICCPR states that freedom of speech carries with it special duties and responsibilities. Therefore, it could be subject to certain restrictions, "but these shall only be such as are provided by law and are necessary: (a) For respect of the rights or reputations of others; (b) For the protection of national security or of public order (*ordre public*), or of public health or morals". According to the HRC, Article 19 of the ICCPR lays down specific conditions for restricting free speech: the restrictions must be "provided by law"; they may only be imposed for one of the grounds specified in the article itself (respect of the rights or reputations of others, the protection of national security or of public, and the protection of public health or morals); and they must conform to the strict tests of necessity and proportionality (CCPR, 2011a). The HRC has emphasized that restrictions on free speech "must not be overboard" (CCPR, 2011a: para. 34).

Generally speaking, the HRC emphasized that restrictions "must conform to the principle of proportionality; they must be appropriate to achieve their protective function; they must be the least intrusive instrument amongst those which might achieve the desired result; and they must be proportionate to the interest to be protected" (CCPR, 1999: para 14). It further added that the principle of proportionality must be respected "not only in the law that frames the restrictions, but also by the administrative and judicial authorities in applying the law" (CCPR, 1999: para 15).

Article 10(2) of the ECHR contains a similar language:

> The exercise of these freedoms, since it carries with it duties and responsibilities, may be subject to such formalities, conditions, restrictions or penalties as are prescribed by law and are necessary in a democratic society, in the interests of national security, territorial integrity or public safety, for the prevention of disorder or crime, for the protection of health or morals, for the protection of the reputation or rights of others, for preventing the disclosure of information received in confidence, or for maintaining the authority and impartiality of the judiciary.

According to the ECtHR, 'necessary' within the meaning of Article 10(2) implies the existence of a pressing social need. Even if such pressing social need exists, the interference in the exercise of free speech must be proportionate to the legitimate aim pursued. For a measure to be necessary and proportionate in a democratic society, there must be no other means of achieving the same end that would interfere less seriously with the fundamental right to freedom of speech (COE, 2020).

2.1 Freedom of Speech and the Right to Access Information

Among the first international human rights institutions to recognize that the right to free speech includes an entitlement to access information held by the authorities is the Inter-American Court of Human Rights (IACtHR). In 2006, the IACtHR held in *Claude Reyes et al. v Chile* that Article 13 of the American Convention on Human

Rights "protects the right of all individuals to request access to State-held information, with the exceptions permitted by the restrictions established in the Convention" (IACtHR, 2006, para 77).

In 2009, the HRC decided in *Mavlonov and Sa'di v Uzbekistan* that "the public has a right to receive information as a corollary of the specific function of a journalist and/or editor to impart information" (CCPR, 2009, para 8.4).

In *Toktakunov v. Kyrgyzstan* the HRC declared that the refusal by the State to provide information on the number of individuals sentenced to death to the legal consultant of a Kyrgyz human rights organization, violates the right to seek and receive information guaranteed by article 19 of the ICCPR. The HRC further held that a legal consultant of a human rights public association can be seen as having a special "watchdog" function on issues of public interest. Finally, it pointed out that certain restrictions on information are permitted only when such restriction are provided by law and are necessary with the aim of respecting the rights or reputations of others, and for the protection of national security or of public order, or of public health or morals. It concluded that restricting the right of the author to access information on the application to the death penalty held by public bodies cannot be deemed necessary for the protection of national security or of public order, public health or morals, or for respect of the rights or reputations of others (CCPR, 2011b).

In its General Comment 34, the HRC recognized that Article 19 of the ICCPR embraces a right of access to information held by public bodies, regardless of the form in which the information is stored, its source and the date of its production (CCPR, 2011a: para. 18). Additionally, the HRC made an explicit reference to the duty of the State to facilitate the enjoyment of the right to access information by "proactively put[ting] in the public domain Government information of public interest" and by ensuring "easy, prompt, effective and practical access to such information" and enact the necessary procedures to facilitate access to information, such as by means of freedom of information legislation (CCPR, 2011a: para 19). Furthermore, the HRC pointed out that the authorities should provide reasons for any refusal to provide access to specific information and should put in place procedures for appealing denial of access to information or failure to respond to requests (CCPR, 2011a: para 19). The HRC further stated that the Article 19 further embraces a right "whereby the media may receive information on the basis of which it can carry out its function" (CCPR, 2011a: para. 13).

While the ECtHR has yet to recognize that Article 10 of the ECHR encompasses a self-standing right of access to State-held information, in *Magyar Helsinki Bizottság v. Hungary* the Grand Chamber of the court reiterated its previous position that such a right may arise in circumstances where access to the information is instrumental for an individual's exercise of their right to freedom of expression. Additionally, the Grand Chamber set out four criteria that could help determining on a case-by-case basis whether a denial of access is detrimental to the exercise of the right to free speech. The first criterion is the purpose of the information requested. According to this criterion the right to access arises under Article 10 of the ECHR when the information sought is necessary for the exercise of the right to freedom of expression. The second criterion is the nature of the information sought. The access sought must generally meet a "public-interest test" for the disclosure to be necessary under Article 10. The third criterion is the role of the applicant. When the individual seeking access to information is doing so with a view to informing the public in a capacity as a public or social watchdog, this will be an important consideration in determining whether Article 10 applies. The Grand Chamber further clarified that this role is not only carried out by the press and civil society but also by bloggers and popular users of the social media, given the important role played by the Internet in enhancing the public's access to news and facilitating the dissemination of information. The forth criterion is the availability and readiness of the information requested (ECtHR, 2016).

2.2 Derogation from the Right to Free Speech

International law contemplates situations in which certain human rights could be suspended temporarily due to exigent circumstances. According to Article 4 of the ICCPR:

> In time of public emergency which threatens the life of the nation and the existence of which is officially proclaimed, the States Parties to the present Covenant may take measures derogating from their obligations under the present Covenant to the extent strictly required by the exigencies of the situation, provided that such measures are not inconsistent with their other obligations under international law and do not involve discrimination solely on the ground of race, colour, sex, language, religion or social origin.

Derogation cannot be used with respect to rights that are considered non-derogable, such as the right to life, the right to be free from torture, the right to be free from slavery or servitude, and the right to be free from retroactive application of penal laws. While COVID-19 pandemic could qualify as a public health emergency, it does not justify derogation from all rights that could be subject to derogation. More importantly, the HRC clarified that even rights that could be subject to derogations in times of emergency, there are certain components of those rights that cannot be made subject to lawful derogation. For example, the right to hold opinions without interference is one such element of Article 19 on free speech that can never be derogated from during a state of emergency (CCPR, 2011a: para. 5).

Furthermore, even when the derogation from certain rights is justified, the principle of proportionality must be respected, as highlighted by the HRC:

> the obligation to limit any derogations to those strictly required by the exigencies of the situation reflects the principle of proportionality which is common to derogation and limitation powers. Moreover, the mere fact that a permissible derogation from a specific provision may, of itself, be justified by the exigencies of the situation does not obviate the requirement that specific measures taken pursuant to the derogation must also be shown to be required by the exigencies of the situation. (CCPR, 2001: para 4)

The HRC further added that States must provide "careful justification not only for their decision to proclaim a state of emergency but also for any specific measures based on such a proclamation" (CCPR, 2001: para 5).

Article 15.1 of the ECHR also allows derogation from certain rights in time of war or other public emergency threatening the life of the nation. Article 15.3 requires States availing themselves of this right of derogation to keep the Secretary General of the Council of Europe fully informed of the measures which it has taken and the reasons therefor (Council of Europe, 1950). The European Commission for Democracy through Law (Venice Commission) (2020) emphasized in a recent report that the only legitimate aim for adopting emergency measures is to help the State overcome the exceptional situation. The nature, severity and duration of the exceptional situation determine the type, extent and duration of the measures that the State may lawfully deploy.

The Venice Commission further emphasized that the principles of necessity and proportionality apply to states of emergency. Necessity entails that only measures which are necessary to help the State overcome the exceptional situation may be justified. Proportionality entails that States are not allowed to resort to measures that are disproportionate to the legitimate aim pursued. When States can choose between several measures, they must choose the ones which are less restrictive (Venice Commission, 2020). More importantly, the Venice Commission emphasized that emergency measures should respect certain general principles of law that could minimize the harm to fundamental rights, democracy and rule of law. Even in a state of public emergency the fundamental principle of the rule of law must prevail. Among the most important aspects of the rule of law that must be maintained in an integral way are: the legality principle; human rights; protection of privacy; democratic participation in and supervision of public decision making; transparency of government; and freedom of expression, association and assembly. The Venice Commission warned against restricting freedom of expression in times of emergency, and hence depriving the public of an essential check on the increased executive powers (The Venice Commission, 2020).

Ten Council of Europe member states declared full or partial states of emergency. Nine countries submitted notifications of derogation from the ECHR, athough none derogated specifically from Article 10 guarantying the right to freedom of expression (Noorlander, 2020). This is not surprising, since the need to gather, circulate and discuss information on the threat of the COVID-19 virus to enable public debates on legitimate differences of expert opinions is essential.

3 Violations in Spain

Three types of violations of the right to free speech were recorded in Spain during the pandemic and in relation to its management. Those are: (a) physical attacks on the journalists covering protests related to the pandemic. Although these attacks were committed by non-State actors, the State is under the obligation to respond to them; (b) lack of transparency and hurdles in accessing information held by authorities; and (c) the use or the threat to use the penal law as a tool to combat disinformation.

3.1 Physical Attack on Journalists

Physical attacks, intimidation and harassments against journalists in relation to their coverage of the pandemic were documented across Europe. According to Partner Organisations of the Platform to Promote the Protection of Journalism and Safety of Journalists (Partner Organizations), such incidents were reported during rallies and demonstrations in France, Greece, Italy, Poland, Russia, Serbia, Turkey, the United Kingdom and Spain (Partner Organisations, 2021). In May 2020, a TVE camera operator was aggressively rebuked by a group of 30 people participating in a protest against the Government in Zaragoza while doing his job. A second incident took place in November, where protesters pushed and insulted a La Sexta TV reporter at a meeting of coronavirus deniers. The same month, in Madrid, a reporter for the newspaper La Razón was also attacked by protesters (PDLI, 2021). While these attacks were carried out by non-State actors, States have the obligation to protect their citizens and those under their jurisdiction from harms inflicted by private actors. States are under the obligation to take appropriate measures or to exercise due diligence to prevent, punish, investigate or redress the harm caused by rights' violations committed by private actors (CCPR, 2004). This obligation also applies to freedom of expression. States are required to ensure that persons are protected from any acts by private persons or entities that would impair the enjoyment of the freedoms of opinion and expression (CCPR, 2011a, para 7). As of the date of finishing this chapter no information was available on legal measures adopted by the authorities to respond to these incidents.

3.2 Lack of Transparency and Hurdles in Accessing Information

During a pandemic, access to information is crucial. In March 2020 the monitors for freedom of expression and freedom of the media for the UN, the Inter-American Commission for Human Rights, and the Representative on Freedom of the Media of the Organization for Security and Co-operation in Europe (OSCE) issued a joint statement on freedom of speech during the pandemic. The statement emphasized that the right to life and the right to health depend on accessing accurate information about the nature of the threats imposed by COVID-19 and on the means to protect oneself, one's family, and one's community from the pandemic. Therefore, it is essential for governments to provide truthful and reliable information, in accessible formats, about the nature of the threat posed by COVID-19. The joint statement urged all governments to robustly implement their freedom of information laws to ensure that all individuals have access to information (OSCE, 2020).

The joint statement focused also on the crucial role of journalism and journalists in times of public health emergency, particularly in informing the public of critical information and monitoring government actions. It called upon governments to make exceptional efforts to protect the work of journalists (OSCE, 2020).

In Spain, the right to access information held by State authorities is set forth in the Act 19/2013 on Transparency, Access to Public Information and Good Governance (2013). This law allows Spanish citizens to request information from public administrations, although in practice, it sets many limits to the exercise of this right and considers the act of requesting information as a mere administrative procedure (Access Info, 2020). With the declaration of the state of alarm by Royal Decree 463/2020 (2020), all administrative procedures were suspended. A few days later, an amendment to Royal Decree 465/2020 (2020) was approved, establishing that the corresponding bodies may decide whether to continue with the administrative procedures related to COVID-19 or not, provided they justify their decision. This amendment released the public administrations from their obligation to process requests for information until the end of the state of alarm, unless they wish to do so (Access Info, 2020). At the regional governance level, substantial disparities existed between the different autonomous communities. Some autonomous communities, such as Castilla y León, Asturias, Castilla-La Mancha and La Rioja, continued to process requests for information, others, like Andalusia, the Canary Islands and Murcia suspended all requests. Some autonomous communities, such as Madrid and Catalonia, processed only part of the requests they received (Access Info, 2020).

Another restriction on freedom of information, implemented at the beginning of the pandemic, was recorded in the government's daily press conferences. Governments in most EU states held regular press conferences, led by senior government figures and health professionals to provide updates regarding the pandemic. However, in Spain questions had to be sent through a WhatsApp, filtered by Spain's secretary of state for communication. This practice was eventually abandoned due to rising public criticism (IPI, 2020).

Another issue concerning transparency is the initial refusal of the Government to disclose the names of the members of the scientific committee advising the government on the pandemic. Only in December 2020, and due to public criticism, the Government announced the names of the experts serving on the Committee (PDLI, 2021).

Generally speaking, limitations on freedom of movement have made it harder for journalists to move around and report about the pandemic. According to the International Press Institute (IPI), journalist in Spain had to face additional hurdles in regions hit hard by the pandemic, such as the autonomous community of Madrid, in which journalist were banned from hospitals, health centers, and from interviewing health personnel at their workstations for "security reasons". Interviews with physicians and health worker had to be conducted over the phone (IPI, 2020). In September 2020, health workers in the community of Madrid received an order from the regional government, forbidding them to speak to the media and urging them "never to act on their own". They were further informed that the regional government must authorize each intervention ahead and hand pick the medical personnel for delivering statements. This protocol was written in 2003 but was dormant till September of 2020, when a physician from Fuenlabrada Hospital made critical statements on Cadena SER radio station (IPI, 2020).

Such limitations were not limited to the Madrid region. For example, photographer Alvaro Calvo, who has been covering the pandemic in Aragón region, was able to enter and photograph an elderly residence home in the region. The photos were distributed internationally by Getty Images. Hours after their publication, he received a judicial injunction from the Aragón government, prohibiting their distribution. Getty refused to withhold publishing his photos internationally, but he agreed to make them inaccessible in Spain (IPI, 2020).

3.3 The Use of Penal Law to Combat Disinformation

In February 2020, the World Health Organization (WHO) announced that the COVID-19 pandemic was accompanied by an "infodemic" of misinformation and disinformation, that, in itself, constituted a serious risk to public health and public action (WHO, 2020). However, the response of States to the challenge imposed by disinformation was not always consistent with international human rights standards on freedom of speech.

In April 2020, the Spanish Minister of Justice announced his intention to pursue legal reforms so that "those who contaminate public opinion in a rude way and without any justification won't get a free pass" (PDLI, 2021).

Additionally, according to news reports, on April 15 2020, an email from the Civil Guard was sent to the different commands urging them to identify false news and fake news "likely to cause social stress and disaffection with government institutions" (Cadena SER, 2020a). The email urged security agents to "identify, study and follow up in relation to the situation created by COVID-19 of disinformation campaigns, as well as publications, hoaxes denying COVID-19 and fake news likely to generate social stress and disaffection with government institutions" (Cadena SER, 2020a). The message urged the units to deliver this data every Friday in order to prepare a monographic report on the cybersecurity activities of the Civil Guard linked to COVID-19, to be discussed in future meetings with the head of the ministerial department. Apparently, the aim of these activities is to inform about and control disinformation, and to search for criminal acts committed online in order to report them to prosecutors and judges (Cadena SER, 2020a).

In the same month, the State Attorney General's Office (Fiscalía General del Estado (FGE)) prepared a report in which it identified a dozen crimes that can be committed by spreading false news. The report was prepared by the Technical Secretariat of the FGE with the aim of providing guidance to Spanish prosecutors who are expected to act in cases of "fake news" related to the pandemic. The crimes enumerated in the document include: Hate speech; discovery and disclosure of secrets in relation to publishing personal information; crimes against moral integrity; the crime of public disorder for those who disseminate alarmist fake news about terrorist attacks or catastrophes; slander and insults; false cure scams for COVID-19; and crimes against the market and consumers (FGE, 2020).

Examples of attempts to use of the penal law as a tool for fighting "fake news" include a complaint filed by members of the political party Unidos Podemos (UP) to the State Attorney General for the posting of a video in which an alleged truck driver shows a warehouse with boxes of sanitary material stored in Spain to be shipped to France. The complaint alleged that the dissemination of this video constitutes a crime against the market, criminalized in Article 284 of the Spanish Penal Code of 1995, for creating the impression of scarcity and thus achieving an increase in the prices of sanitary material (Newtral, 2020). Another complaint logged by UP members referred to a video in which coffins can be seen in a huge warehouse, accompanied in the upper part by the acronym PSOE (the governing political party). While the video suggested that these deaths are related to COVID-19 in Spain, these images were taken outside Spain. The complaint alleged that this video constitutes defamation of the institutions of the State, criminalized in Article 504 of the Spanish Penal Code of 1995, for calling the Government of Spain a "son of a bitch" and a "murderer" (Newtral, 2020). While these complaints were eventually closed by the prosecution's office (Libertad Digital, 2020), this type of surveillance and the mere threat of resorting to criminal law to combat "fake news" or disinformation could have a chilling effect on freedom of speech and could lead to self-censorship. As highlighted by the ECtHR (2004, 2015a and 2015b), interferences with the right to freedom of expression arise not only when sanctions are actually imposed, but also from the fear of sanctions and the broader legal and regulatory climate for journalists and speech in general.

Even before the pandemic, human rights experts representing international organizations condemned governments, in a "Joint Declaration on Freedom of Expression and "Fake New", Disinformation and Propaganda", for using vague and ambiguous terms to outlaw the dissemination of certain information, such as the use of the term "false [and] non-objective information" (OSCE, 2017).

During the pandemic, human rights experts representing international organizations issued a joint declaration of freedom of speech and the flow of information during the pandemic (OSCE, 2020). While they shared the concern of governments over the impact of false news on the pandemic, the international human rights experts believed that combating disinformation should be handled primarily by governments and internet companies, by providing reliable information themselves. They also emphasized that measures, such as content takedowns and censorship, may result in limiting access to important information for public health, and could be justified only when the standards of necessity and proportionality are met. As for the attempt to criminalize information, the international human rights experts believed that this could create distrust in institutional information, delay access to reliable information and have a chilling effect on freedom of expression (OSCE, 2020).

It is worth noting that the Spanish national cyber plan follows similar lines by stating that tacking information could be achieved by promoting "a critical spirit in favor of truthful and quality information that contributes to the identification of false news and misinformation" (Council of National Security, 2019: Line of action 7.5).

4 Structural Issues

As mentioned in the introduction, even States with a strong record on human rights failed to meet all of their international human rights obligations related to free speech during the pandemic. The lax protection of free speech during times of crisis is usually made easier when the domestic legal regime for the protection of free speech suffers from certain deficiencies that precede the situation of emergency. In the case of Spain, previous deficiencies that had been repeatedly highlighted by international institutions made it easier and more tempting for the authorities to impose undue restrictions on free speech during the pandemic to minimize public criticism of the management of the pandemic. The following is a discussion of two pre-existing conditions that had a negative impact on freedom of speech during the pandemic.

4.1 Freedom to Access Information not a Fundamental Right

One of the structural legal problems in Spain in relation to the right to access information held by the authorities is the absence of recognition of this right as a fundamental right. In other words, the right to access information held by the authorities is not seen as embedded in article 20 of the Spanish Constitution of 1978, which protects freedom of expression (Access info, 2020). As mentioned earlier, the Law on Transparency, Access to Public Information and Good Governance allows Spanish citizens to request information held by public administrations; however, the act of requesting information is considered a mere administrative procedure (Access Info, 2020).

In 2016, over fifty top Spanish constitutional experts, lawyers, and academics signed a letter calling upon the Spanish government to recognize access to information as a fundamental right in line with international jurisprudence, which links the right to information to freedom of expression. The signatories insisted that there is no need for a constitutional reform as Article 20 of the Spanish Constitution of 1987 already establishes the right to "freely communicate or receive accurate information by any means of dissemination whatsoever" (Article 20.1.d) and the right to "freely express and disseminate thoughts, ideas and opinions trough words, in writing or by any other means of communication" (Article 20.1.a). The signatories emphasized that the failure to recognize the right to access information held by the Government as a fundamental rights goes against the general trend of international and European law and violates Article 10.2 of the Spanish Constitution, which establishes that the norms relating to fundamental rights and freedoms recognized by the Constitution must be interpreted in accordance with the Universal Declaration of Human Rights and international human rights treaties ratified by Spain (Romero et al., 2016).

4.2 Excessive Limitations on Political Criticism

Human rights and civil society groups have long expressed their concern over the use of penal law to unduly restrict freedom of expression in Spain. For example, in their joint submission to the Universal Periodic Review of Spain, held by the UN Human Rights Council, Article 19 and the European Centre for Press and Media Freedom (ECPMF) expressed their concern over the definition of the crime of "insult", which includes the crime of "Lèse majesté" and "insult" of the State, public bodies and State symbols. They also criticized the crime of insulting religious feelings, the crime of incitement and glorification of terrorism, and hate speech (Article 19 & ECPMF, 2019). This chapter focuses only on provisions that could be utilized to punish disinformation containing criticism of the State's and of its officials' handling of the pandemic.

The Spanish Penal Code of 1995 contains three different "lèse majesté" provisions/The first article is **Article 490(3)** of the Spanish Penal Code, which criminalizes "slander" and "insult" against members of the Spanish Royal Family, during or related to the exercise of their official functions. The maximum penalty of this crime is imprisonment of six months to two years if the slander or defamation are serious. The second article is **Article 491(1)** of the penal code, which criminalizes "slander" and "insults" against various members of the Spanish Royal Family, without a connection to the exercise of their official functions. The penalty for this crime is a fine from four to twenty months, based on the day-fine system applicable in Spain. According to Article 50 of the Spanish Penal Code, "the daily quota [of the fine] shall be a minimum of two and a maximum of four hundred euros". Finally, **Article 491(2)** of the penal code criminalizes the use of images of past, present or future Kings or Queens, or other present members of the Royal Family, "in any way that could damage the prestige of the Crown". The penalty of this crime is a fine from six to twenty- four months.

Such penal provisions do not meet international standards on freedom of speech. First, they do not meet the legality test since they are drafted in a broad sense that opens the door for political abuse and arbitrary application. This stands in contradiction with the maxim *nullum crimen sine lege*. Embedded in this maxim is the requirement of *lex certa*, meaning that the law be clear and certain, and the requirement of *lex stricta*, meaning that the law be narrowly construed (Gallant, 2009). Second, such provisions impose unnecessary and disproportionate restrictions on the right to freedom of expression, especially when it comes to criticizing individuals who carry out official duties.

As for the provisions that criminalize insults of state institutions, the penal code contains four different provisions: **Article 496** criminalizes serious defamation against the Parliament or a Legislative Assembly of an Autonomous Community, while in session, or of any of its commissions. The penalty for this crime is a fine from twelve to eighteen months; **Article 504(1)** criminalizes "slander, defamation or threats" against the National Government, the General Council of the Judiciary, the Constitutional Court, the Supreme Court, or the Governing Council or High Court of Justice of an autonomous community. Conviction of this crime carries with it a punishment of a fine from twelve to eighteen months; **Article 504(2)** criminalizes "serious insults" against the Armed Forces and the different security forces and bodies. The penalty for this crime is a fine from twelve to eighteen months; finally, **Article 543** criminalizes verbal or written offences or outrages, or those committed by action, against Spain, its Autonomous Communities or its symbols or emblems, when committed publicly. The penalty for this crime is a fine from seven to twelve months.

It is worth mentioning that in Spain, insulting State institutions and symbols carries with it heavier penalties compared to the defamation of private citizens. Article 208 of the penal code defines defamation as an "action or expression that harms the dignity of another person, detracting from his reputation or attacking his self-esteem. Only defamation that, due to its nature, effects and circumstances, is considered serious by the public at large, shall be deemed to constitute a felony". According to Article 209 of the penal code "severe defamation perpetrated with publicity shall be punished with the penalty of a fine from six to fourteen months and, otherwise, with that of three to seven months".

Attaching higher penalties for the defamation of State institutions compared to the defamation of regular citizens is very telling. The HRC has expressed a deep concern over laws on such matters as, lese majesty, *desacato*, disrespect for authority, disrespect for flags and symbols, defamation of the head of state and the protection of the honor of public officials, as well as laws that provide for more severe penalties solely on the basis of the identity of the person that may have been impugned. It further highlighted that States "should not prohibit criticism of institutions, such as the army or the administration" (CCPR, 2011a: para 38). Furthermore, it highlighted that "with regard to comments about public figures, consideration should be given to avoiding penalizing or otherwise rendering unlawful untrue statements that have been published in error but without malice… a public interest in the subject matter of the criticism should be recognized as a defence" (CCPR, 2011a: para 47).

Generally speaking, the HRC asked States to "consider the decriminalization of defamation and, in any case, the application of the criminal law should only be countenanced in the most serious of cases and imprisonment is never an appropriate penalty" (CCPR, 2011a: para 47). In *Adimayo M. Aduayom, Sofianou T. Diasso and Yawo S. Dobou v. Togo* the HRC emphasized that "citizens must be allowed to inform themselves about alternatives to the political system/parties in power, and that they may criticize or openly and publicly evaluate their governments without fear or interference or punishment" (CCPR, 1996: para 7.4). In *Bodrožić v. Serbia and Montenegro* the HRC dealt with the conviction of a journalist of criminal insult for accusing a Socialist Party leader of corruption, even though a domestic court found that the factual basis of the article was true. The HRC found a violation of article 19 of the ICCPR emphasizing that "in circumstances of public debate in a democratic society, especially in the media, concerning figures in the political domain, the value placed by the Covenant upon uninhibited expression is particularly high" (CCPR, 2005: para. 7.2). This indicates that the HRC sees no necessity, under Article 19(3) of the ICCPR, to protect the rights or reputations of public officials when the accusations leveled against them have a good-faith basis (Carter, 2016). Even if good-faith basis could not be established in relation to accusations against government officials, the HRC had emphasized that the burden is on the state to prove restrictions on freedom of expression are necessary (Carter, 2016). In *Dissanayake, Mudiyanselage Sumanaweera Banda v. Sri Lanka* the HRC held that the imposition of two-years rigorous imprisonment for contempt of court by the Sri Lankan Supreme Court for questioning the independence and integrity of the court, constitutes, inter alia, a violation of Article 19 of the ICCPR. The HRC affirmed that the sentence imposed upon the author was disproportionate to any legitimate aim under article 19, paragraph 3 (CCPR, 2008).

In *Stern Taulats and Roura Capellera v. Spain*, the ECtHR (2018) held that the conviction of two Spanish nationals for setting fire to a photograph of the royal couple at a public demonstration had violated the right to freedom of expression. In *Otegi Mondragon v. Spain* the ECtHR (2011) emphasized that there is a "little scope under Article 10 § 2 for restrictions on freedom of expression in the area of political speech or debate—where freedom of expression is of the utmost importance—or in matters of public interest" (para. 50). It further emphasized that "limits of acceptable criticism are wider as regards a politician as such than as regards a private individual. Unlike the latter, the former inevitably and knowingly lays himself open to close scrutiny of his every word and deed by both journalists and the public at large, and he must consequently display a greater degree of tolerance" (ECtHR, 2011: para 50).

The ECtHR (2011) further clarified that "while any individual who takes part in a public debate of general concern … must not overstep certain limits, particularly with regard to respect for the reputation and rights of others, a degree of exaggeration, or even provocation, is permitted; in other words, a degree of immoderation is allowed" (para 54). The ECtHR concluded that "convicting the applicant based on Article 490 § 3 of the Criminal Code, which affords the Head of State a greater degree of protection than other persons (protected by the ordinary law on insults) or institutions (such as the government and Parliament) … and which lays down heavier penalties for insulting statements … will not, as a rule, be in keeping with the spirit of the Convention" (ECtHR, 2011: para 55).

In 2004, the Committee of Ministers of the COE adopted the Declaration on Freedom of Political debate in the Media (COE, 2004). The declaration states that the "state, the government or any other institution of the executive, legislative or judicial branch may be subject to criticism in the media. Because of their dominant position, these institutions as such should not be protected by criminal law against defamatory or insulting statements" (Article II). The declaration further states that "political figures should not enjoy greater protection of their reputation and other rights than other individuals, and thus more severe sanctions should not be pronounced under domestic law against the media where the latter criticise political figures" (Article VI). Furthermore, the Parliamentary Assembly of the COE adopted Resolution 1577, entitled "Towards decriminalisation of defamation" (COE, 2007). The resolution included the following recommendations for states: to abolish prison sentences for defamation without delay; to guarantee that there is no misuse of criminal prosecutions for defamation; to define defamation more precisely in order to avoid an arbitrary application of the law; to ensure that civil law provides effective protection for persons affected by defamation; and to remove from their defamation laws any increased protection for public figures (COE, 2007).

On March 11, 2021, the Council of Europe Commissioner for Human Rights, Dunja Mijatović sent a letter to the Minister of Justice of Spain inviting the Spanish authorities to amend in a comprehensive fashion penal provisions that contravene freedom of speech and to strengthen existing safeguards in full line with the ECHR. Among the concerns highlighted in her letter is the growing number of criminal convictions based on provisions criminalizing libels and insults to the Crown. The Commissioner emphasized that the possibility of imposing restrictions on freedom of expression in the context of political debate are very limited, in particular when it comes to politicians, representatives of the authorities and other public figures (Mijatović, 2021).

Another legal challenge to freedom of speech in Spain is the use of the term "hate speech" to outlaw political criticism of people holding public office. For example, in the above-mentioned case of Stern *Taulats and Roura Capellera v. Spain,* the ECtHR criticized the ruling of the Spanish Constitutional Court on the conviction of the applicants, according to which "the public burning the portrait of the Monarchs is an act not only offensive but also it incites to hatred" (Quoted in ECtHR, 2011: para 14). The ECtHR further emphasized that criticism of public authorities and personalities, as provocative and radical as it may be, could not be considered as hate speech and incitement to violence (ECtHR, 2011). The ECtHR reaffirmed its previous position that in a democratic, pluralistic and tolerant society, the right to free speech protects also expressions that offend, shock or disturb the State or any sector of the population. It also referred to Recommendation No. R (97) 20 of the Committee of Ministers of the COE on hate speech, which defines the latter as "all forms of expression which spread, incite, promote or justify racial hatred, xenophobia, anti-Semitism or other forms of hatred based on intolerance, including: intolerance expressed by aggressive nationalism and ethnocentrism, discrimination and hostility against minorities, migrants and people of immigrant origin" (COE, 1997).

Also, the letter of the Council of Europe Commissioner for Human Rights called upon the Spanish authorities to restrict the application of provisions related to hate speech to cases prohibited under international human rights law, that is to expressions of hatred based on xenophobia, antisemitism and other forms of hatred based on intolerance that constitute incitement to discrimination, hostility or violence ((Mijatović, 2021). However, just recently it was reported that a retired 70-year-old teacher was indicted with the crime of hate speech for a comment she made on Facebook against a Civil Guard agent who participated in the operation to prevent the 2017 referendum in Catalonia (Diario, 2021).

5 Conclusions

Freedom of speech is a basic human right protected by international and regional human rights treaties. As mentioned earlier, freedom of speech is essential for maintaining a democratic order and for the protection of a wide spectrum of other human rights. Even with its centrality, free speech remains one of the most violated human rights on a global scale. Paradoxically, when freedom of speech was more crucial and necessary, it was subject to intensified attacks during the COVID-19 with the aim of silencing criticism of governmental management of this global crisis. Even in democratic counties with a strong record on freedom of speech, violations of international standards of free speech were not uncommon. In Spain, three types of violations of the right to free speech were recorded: (a) physical attacks on the journalists covering protests related to the pandemic; (b) lack of transparency and hurdles in accessing information held by authorities; and (c) the use or the treat to use the penal law as a tool to combat disinformation in relation to the pandemic. This violation, at least the one concerning lack of access to information and the use of penal tools to combat disinformation could be attributed to deficiencies in the protection of free speech that existed prior to the pandemic and that made it easier for the authorities to unduly restrict freedom of speech in times when such rights had a crucial importance for informing the public on the serious health threat that they were facing. Such legal gaps made it easier for the authorities to evade their responsibility in providing timely and accurate information on the pandemic. They also hindered the function of journalist in reporting freely on emergency measures adopted by the government on the pandemic and hence had the potential of depriving the public of an essential check on the increased executive powers assumed by governments.

These legal gaps made it tempting for the authorities to threaten with the use of penal sanctions to minimize criticism of the government under the pretext of fighting disinformation in relation to the pandemic. When asked about measures to combat disinformation, José Manuel Santiago, Chief of the General Staff of the Civil Guard, stated that one line of action is to "minimize this climate contrary to crisis management by the Government" (Cadena SER, 2020b). It is worth noting that the high-level group of experts, convened by the European Commission, defined disinformation as false, inaccurate, or misleading information designed, presented and promoted to intentionally cause public harm. Public harm was framed in terms of "threats to democratic political processes and values" (European Commission, 2018: 10). This is very different from the mere criticism of a specific government or specific institutions for their management of a crisis.

This is a clear reminder on how legal gaps in protection of freedom of speech could open the door for far-reaching limitation of freedom of speech in times of emergency when freedom of information and free press are more needed than ever. This could happen also in vibrant democracies like Spain. Needless to say, the dangers of such excessive restrictions on freedom of speech could have wider implication for the future, even after the emergency is over. The limitations imposed during the pandemic run the risk of it creating an environment that is detrimental to public criticism of state's institutions and politicians, and environment that have a chilling effect on freedom of speech, and that leads to self-censorship by individuals and the media alike.

References

Access Info. (2020). *Spain must guarantee right to access information during coronavirus crisis*. https://www.liberties.eu/en/stories/spain-must-guarantee-right-access-information-during-coronavirus-crisis/18651

Act 19/2013 on Transparency, Access to Public Information and Good Governance. BOE-A-2013-12887.

Alonso Romero, E., et al. (2016 December 10). Alonso Romero et al. to the Spanish Government. *For the recognition of the right to access information as a fundamental rights*. 2016 December 10, (In Spanish). https://www.access-info.org/wp-content/uploads/Declaraci%C3%B3n-Dcho-Informaci%C3%B3n-Fundamental-Acad%C3%A9micos.docx

Article 19 & ECPMF. (2019). Submission to the universal periodic review of Spain by Article 19 and the European Centre for Press and Media Freedom (ECPMF) for consideration at the 35th session of the working group in January 2020.

Article 19. (2021). The Global Expression Report 2021. https://www.article19.org/wp-content/uploads/2021/07/A19-GxR-2021-FINAL.pdf

Cadena SER. (2020a). Guardia Civil insta a identificar fake news susceptibles de crear "desafección a instituciones del Gobierno". https://cadenaser.com/ser/2020/04/20/tribunales/1587410447_579580.html

Cadena SER. (2020b). Polémica por las palabras del Jefe del Estado Mayor de la Guardia Civil sobre "minimizar" críticas al Gobierno. https://cadenaser.com/ser/2020/04/19/tribunales/1587295094_340879.html

Carter, E. L. (2016). "Error but without malice" in defamation of public officials: The value of free expression in international human rights law. *Communication Law and Policy, 21*(3), 301–322.

CCPR. (1996). *Adimayo M. Aduayom, Sofianou T. Diasso and Yawo S. Dobou v. Togo*. Un Doc. CCPR/C/51/D/422/1990, 423/1990 and 424/1990.

CCPR. (1999). *General Comment No. 27. Freedom of movement (article 12)*. UN Doc. CCPR/C/21/Rev.1/Add.9.

CCPR. (2001). *General comment no. 29 states of emergency (article 4)*. UN Doc. CCPR/C/21/Rev.1/Add.11.

CCPR. (2004). *General Comment No. 31. The nature of the general legal obligation imposed on states parties to the Covenant*. UN Doc. CCPR/C/21/Rev.1/Add. 1326.

CCPR. (2005). *Bodrožić v. Serbia and Montenegro*. UN Doc. CCPR/C/85/D/1180/200.

CCPR. (2008). *Dissanayake, Mudiyanselage Sumanaweera Banda v. Sri Lanka*. UN Doc. CCPR/C/93/D/1373/2005.

CCPR. (2009). *Rakhim Mavlonov and Shansiy Sa'di v. Uzbekistan*, UN Doc. CCPR/C/95/D/1334/2004.

CCPR. (2011a). *General comment No. 34. Article 19: Freedoms of opinion and expression*. UN Doc. CCPR/C/GC/34.

CCPR. (2011b). *Toktakunov v. Kyrgyzstan*. UN Doc. CCPR/C/101/D/1470/2006.

COE. (1950). *European convention for the protection of human rights and fundamental freedoms, as amended by protocols Nos. 11 and 14, 4 November 1950, ETS 5*. https://www.refworld.org/docid/3ae6b3b04.html

COE. (1997). *Recommendation No. R (97) 20 of the Committee of Ministers of the Council of Europe on Hate Speech*. https://search.coe.int/cm/Pages/result_details.aspx?ObjectID=0900001680505d5b

COE. (2004). *Declaration on freedom of political debate in the media* (Adopted by the Committee of Ministers on 12 February 2004). https://search.coe.int/cm/Pages/result_details.aspx?ObjectId=09000016805dddf8

COE. (2007). *Resolution 1577. Towards decriminalisation of defamation.* Parliamentary Assembly. https://assembly.coe.int/nw/xml/XRef/Xref-XML2HTML-en.asp?fileid=11684&lang=en

COE. (2020). *Guide on Article 10 of the European Convention on Human Rights.* https://www.echr.coe.int/documents/guide_art_10_eng.pdf

Council of National Security. (2019). *National cybersecurity strategy 2019.* https://www.boe.es/boe/dias/2019/04/30/pdfs/BOE-A-2019-6347.pdf

Diario, E. (2021). *Una jueza imputa por delito de odio a una jubilada por un comentario en Facebook contra un guardia civil.* https://www.eldiario.es/catalunya/jueza-imputa-delito-odio-profesora-jubilada-comentario-facebook-guardia-civil_1_8173019.html

ECtHR. (1976). Handyside v. United Kingdom, Application No. 5493/72.

ECtHR. (2004). Pedersen and Baadsgaard v. Denmark, Application No. 49017/99.

ECtHR. (2011). Otegi Mondragon v. Spain, Application no. 2034/07.

ECtHR. (2015a). Yaşar Kaplan v. Turkey, Application No. 56566/00.

ECtHR. (2015b). Pentikäinen v. Finland, Case No. 11882/10.

ECtHR. (2016). Magyar Helsinki Bizottság v. Hungary, Application no. 18030/11.

ECtHR. (2018). Stern Taulats and Roura Capellera v. Spain, Application No. 51168/15.

European Commission. (2018). *A multi-dimensional approach to disinformation.* Report of the independent High level Group on fake news and online disinformation. JOIN (2018) 36 final.

FGE. (2020). *Tratamiento Penal de las "Fake News".* https://www.icab.es/export/sites/icab/.galleries/documents-noticies/tratamiento-penal-de-las-fake-news-fiscalia-general-del-estado.pdf

Gallant, K. S. (2009). *The principle of legality in international and comparative criminal law.* Cambridge University Press.

HRW. (2021). *Covid-19 triggers wave of free speech abuse* (2021 February 11). Covid-19 Triggers Wave of Free Speech Abuse|Human Rights Watch (hrw.org)

IACtHR. (1985). *Compulsory membership in an association prescribed by law for the practice of journalism (Arts 13 and 29 American Convention on Human Rights), Advisory Opinion OC-5/85, Series A, No 5.*

IACtHR. (2006). Case of Claude Reyes et al v Chile, Claude Reyes et al. v Chile, Merits, reparations and costs, IACHR Series C no 151, IHRL 1535.

IPI. (2020). *Spain's free press put to test under COVID-19 restrictions.* https://ipi.media/spains-free-press-put-to-test-under-covid-19-restrictions/

Libertad Digital. (2020). *La Fiscalía archiva la investigación por difundir bulos sobre la covid-19 en redes sociales que denunció Podemos.* https://www.libertaddigital.com/espana/2020-09-04/fiscalia-archiva-investigacion-bulos-covid-19-podemos-1276663464/

McGoldrick, D. (2017). Thought, expression, association, and assembly. In S. Shah & S. S. D. Moeckli (Eds.), *International human rights law* (pp. 208–232). Oxford University Press.

Mijatović, D. (2021 March 21). *Dunja Mijatović, the Council of Europe Commissioner for Human Rights, to Juan Carlos Campo, the Minister of Justice of Spain.* https://rm.coe.int/letter-to-mr-mr-juan-carlos-campo-minister-of-justice-of-spain-by-dunj/1680a1c05e

Newtral. (2020). *De WhatsApp al Código Penal: cuando un bulo se puede convertir en delito.* https://www.newtral.es/de-whatsapp-al-codigo-penal-cuando-un-bulo-se-puede-convertir-en-delito/20200407/

Noorlander, P. (2020). COVID and free speech. In *Council of Europe, Background Paper, Ministerial Conference, Cyprus.*

O'Flaherty, M. (2012). Freedom of expression: Article 19 of the international covenant on civil and political rights and the human rights committee's general comment no 34. *Human Rights Law Review, 12*(4), 627–654.

OSCE. (2017). *Joint declaration on freedom of expression and "fake new", disinformation and propaganda.* https://www.osce.org/files/f/documents/6/8/302796.pdf

OSCE. (2020). *COVID-19: Governments must promote and protect access to and free flow of information during pandemic, say international media freedom experts.* https://www.osce.org/representative-on-freedom-of-media/448849
Partner Organisations. (2021). *Annual report by the partner organisations to the council of Europe platform to promote the protection of journalism and safety of journalists.* Safety of journalist platform.
PDLI. (2021). *2020: Otro Año Para La Libertad de Expresión en España.* http://libertadinformacion.cc/2020-otro-mal-ano-para-la-libertad-de-expresion-en-espana/
Royal Decree 463/2020, of March 14th, 2020, Declaring the State of alarm in Spain to manage the health crisis situation caused by COVID-19. BOE-A-2020-3692.
Royal Decree 465/2020, of 17 March, which modifies Royal Decree 463/2020, of 14 March, Declaring the state of alarm in Spain to manage the health crisis situation caused by COVID-19. BOE-A-2020-3828.
The Spanish Constitution of (1987). https://www.boe.es/legislacion/documentos/ConstitucionINGLES.pdf
The Spanish Penal Code. (Organic Law No. 10/1995 of November 23, (1995), as amended up to Organic Law No. 2/2019 of March 1, 2019). https://wipolex.wipo.int/en/legislation/details/18760
UN. (1946). *Calling of an international conference on freedom of information.* UN Doc. A/RES/59.
UN. (1966). International covenant on civil and political rights. *Treaty Series* (p. 171), *999.*
Venice Commission. (2020). *Respect for democracy, human rights and the rule of law during states of emergency-reflectins.* CDL-PI(2020)005rev.
WHO. (2020). *Novel coronavirus (2019-nCoV).* Situation Report—13. https://www.who.int/docs/default-source/coronaviruse/situation-reports/20200202-sitrep-13-ncov-v3.pdf

Sonia Boulos holds a Doctorate Degree in Juridical Science (JSD) from the University of Notre Dame (USA). Currently, she works as a professor of law at the department of International Relations, Nebrija University. Her teaching and research focus on the international protection of human rights. She has authored several publications in this field, including in high-impact journals, such as the Journal of Business ethics.

Media Securitization in the Migration Crisis in Spain 2020–2021: The Case of Canary Islands and Ceuta

José-Manuel Moreno-Mercado, Adolfo Calatrava-García, and José-Miguel Calvillo-Cisneros

Abstract Migrations are positioned as a security issue in a context marked by multidimensionality. One of the derivatives of the extension of the concept of security, which is evident in the case of migration policies, is the process of securitization. This concept, derived from constructivist approaches to international relations and security, particularly the Copenhagen School, explains how the political problems defined by the establishment of public policies end up in the field of security by the speech act of the main actors. In this sense, the theories of securitization are increasingly being studied in political communication research. This proposed chapter aims to measure the presence of security and human drama frames in the Spanish press during two specific migration crises; The crises of the Canary Islands (2020) and Ceuta (2021). We hypothesize that the press has undergone a process of media securitization as it happens in the coverage of armed conflicts. For this, we will resort to text mining and Natural Language Processing (NLP) to study large volumes of information. This proposal, therefore, seeks to have a dual dimension: theoretical and methodological.

Keywords Migration · Securitization · Crisis · Framing · Press · Communication

1 Introduction: Migration as a Media Securitization Issue

Migration policies in Europe have taken on an extraordinary dimension (Arango, 2007). The end of the Cold War and, above all, the terrorist attacks of September 11,

J.-M. Moreno-Mercado (✉)
Carretera de La Sierra, n.º 94, 1º A. CP: 18008, Granada, Spain
e-mail: jmmoreno95@correo.ugr.es

A. Calatrava-García · J.-M. Calvillo-Cisneros
Campus Somosaguas., CP: 28223, Madrid, Pozuelo de Alarcón, Spain
e-mail: acalatra@ucm.es

J.-M. Calvillo-Cisneros
e-mail: jcalvill@ucm.es

J. Cayón Peña (ed.), *Security and Defence: Ethical and Legal Challenges in the Face of Current Conflicts*, Advanced Sciences and Technologies for Security Applications,
https://doi.org/10.1007/978-3-030-95939-5_15

2001, have brought immigration into the security policy arena. Since these two historical milestones, international migration has become a complex phenomenon and part of a deeply interconnected world. These milestones were the push for migration to have a security dimension. As a result, migration issues have been incorporated into the security agendas of governments and international organizations. In European migration policy, conflicts in the Middle East, North Africa, and the Sahel have accelerated the prominence of migration in these agendas. In this sense, Europe is a destination for migratory routes from Africa and the Middle East which cross the Mediterranean in search of a safer life.

In this way, countries of origin, transit routes, and host countries are interconnected systems where changes in any of the three routes: eastern, central, and western can have consequences. In this new scenario, most international migration takes place legally –fully respecting the legal frameworks- of the receiving states. However, governments are focused on illegal migration and mafias.

Spain has developed from 2006 a migration policy which has had continuity throughout different governments. First, a socialist government created the migration policy in the context of the cayuco crisis. However, it was later further developed by both right and left governments. The cayuco crisis implied the arrival of more than thirty thousand migrants during 2006 to the coasts of the Canary Islands in boats (known as cayucos). Since then, migration policies have undergone a process of politicization and securitization. This process has been determined by two frames related to the migrants themselves: human rights and security.

In this chapter, we want to study these frames in the media. To do so, we will study two "migration crises" in recent months: one at the end of 2020 in the Canary Islands and the other in mid-2021 in Ceuta. The term migration crisis is usually used to refer to the massive and uncontrolled arrival of migrants. The article aims to study the frames mentioned in the Spanish press and Morocco's role in the media narratives.

2 The Migration-Security Nexus. A Theoretical Debate Without Empirical Evidence

The end of the Cold War brought about a renewal of the debate on threats to the security of States. Traditionally, the classical theoretical schools of International Relations have given great importance to threats from other states. This vision is from and for states and does not consider the rest of the actors in international society. International society has not changed much because states and power continue to be the central elements in international relations.

However, the new non-governmental actors acting in international space are conditioning the behavior of states. Therefore, we can affirm that: "International society is an organization defined by one principle: the interest of the state and its security. States are always seeking their survival and, therefore, security takes on a dimension which surpasses all other issues" (Waltz, 1979). Similarly, Keohane and

Nye's (1989) theory of complex interdependence recognizes the existence of multiple channels connecting societies, including informal links between governmental elites and informal links between non-governmental elites and transnational organizations. These channels are located at the three levels of analysis of International Relations: interstate, trans-governmental and transnational. According to "complex interdependence", the agenda of international relations is made up of various issues without a clear hierarchy, which means that military security no longer dominates the agenda.

Alexander Wendt (1999) establishes that international politics is fundamentally a social construction from the Critical International Relations Theory. This school criticizes traditional ways of understanding international politics and presents itself as a theoretical alternative to rationalist approaches. However, Wendt agrees that the causes threatening the security of states have multiplied. Wendt (critical) sees it as a process of social construction, and Waltz finds the explanation in the disorder of international society and the emergence of new actors.

We can admit that security is still at the center of all strategic analyses. However, since the 1990s, security has taken on a multidimensional character, moving away from the classic Cold War security concept.

The new redefinition of the concept gave much importance to new dimensions and threats to state sovereignty. Some of them were migration, fundamentalism, hunger, development, climate change, and natural disasters. All these issues create the global agenda or the new world political agenda. In this sense, Buzan et al. (1998) establish a holistic definition of security. According to this definition, in addition to the more traditional military dimension of security, there is a political dimension, including authority relations, internal legitimacy, and internal stability. It also includes an economic dimension, covering trade and production relations, access to resources and markets; a social dimension, which includes the identity relations of groups and communities; and an environmental dimension, which establishes relations between human beings and the biosphere.

In this way, we can position migrations as a security issue within a multidimensional space where non-state actors play a more critical role. Thus, current debates on migration and security reflect changes like migration and the way of thinking about it (Huysmans and Squire, 2009). Current studies on international migration are included under security studies. In addition, states and international organizations introduce this perspective in their internal and external security strategies.

In migration policies, securitization processes are one of the extensions of the concept of security. In this way, the spoken acts of the main actors create a narrative and a new public policy. In addition, securitization processes allow elites to control these issues because they have constructed a negative perception of migrations.

Considering a specific issue as a security problem limits the political or ideological debate since it is based on considerations which affect national interests. Thus, the problem is determining whether a political problem is a security issue (Abad-Quintanal, 2015: 50) but, once achieved, the design of a security policy is justified.

In Europe and the United States, the immigration policies' politicization has been a previous element in the securitization process. For example, in the electoral processes since the middle of the last decade, migration has been one of the star issues in many European countries and the two presidential elections in the United States (2016 and 2020) (Arango et al., 2019).

In the 2003 European Security Strategy (ESS), cross-border immigrant smuggling is linked to organized crime. Moreover, in Spain, the 2003 Strategic Defense Review (RED) indicated that one of the risks to national security was uncontrolled migratory movements. Therefore, in Spain's three national security strategies of 2011, 2013, and 2017, irregular migratory flows have been part of the risks and challenges to national security.

In conclusion, migration policy should balance security issues, reception and integration policies, and respect for human rights. There have been repeated images and speeches linking migration with negative aspects during the last three decades, but labeling immigrants as a "threat" or "security problem" is simplistic and has no empirical evidence to support it. This type of linkage without supporting evidence influences the rise of xenophobia and anti-immigrant sentiments, especially in developed countries, where the immigrant becomes a threat to economic and social stability.

In conclusion, there are three different theoretical approaches linking migration and security, which explicitly combine internal and external state security elements: (1) The relationship between national security, border control, and migration, which indicates that the State must regulate these flows. (2) Organized crime and human traffickers are presented as a new threat enhanced by globalization. (3) Issues related to the integration of migrants, associated with elements of identity and linked to issues such as radicalization and terrorism.

3 Method and Research Design

The research hypothesis is based on the theoretical postulates set out above. That is, the presence of the security frame as a generic journalistic routine of the Spanish press. The following hypothesis is formulated:

> H1: The security frame has a majority use by the Spanish press when reporting on the migration crises in the Canary Islands, Ceuta and Melilla.

To verify or refute the hypothesis, the news of the main press newspapers in Spain have been analyzed, specifically *ABC*, *El Mundo*, *El País* and *La Vanguardia*. These four media outlets are of great national circulation and of divergent ideological lines. However, the construction of the hypothesis starts from the premise that the security frame will be dominant in all media despite its ideological spectrum. This is because the Spanish press has experienced a progressive process of media securitization when it comes to reporting on international crises, as in the case of the war in Syria (Moreno-Mercado et al., 2021). Although, it should be noted that there is no academic unanimity on how the media in Spain report on immigration. On the one hand, several scholars point out that journalists in Spain tend to defend the human rights of immigrants by showing the positive effects of immigration (Oller-Alonso et al., 2021). And, on the other hand, various studies indicate that there is a certain de-citizenship of immigrants (Fajardo-Fernández & Soriano-Miras, 2016).

The study period includes one-month coverage of both migration crises. From 1 November 2020 to 1 December 2020 in the case of the Canary Islands and from 17 April 2021 to 17 June 2021 in the case of Ceuta and Melilla. Both time periods encompass the key moments of both migration crises and allow us to work with a considerable sample of news. The news has been extracted from the MyNews database with the following search terms: "*Crisis*", "*Canary Islands*", "*Ceuta*" and "*Melilla*". Keywords could be found in either the headline, subtitle or body of the news. The total sample of the study comprises 877 news stories.

In short, for the demonstration of the hypothesis, the presence of security and human rights frames has been measured. These frames can be described according to the categories of Entman (1993). The frames are not mutually exclusive although they can traditionally be. The security frame emphasizes geopolitical issues and immigration as a threat which can create instability. On the contrary, the human rights frame defines the problem of immigration as a humanitarian drama, focusing on more emotional aspects. In summary, the research focuses on two generic frames typical of international crises.

Once the coding book has been prepared, both the headlines and the body of the news have been preprocessed. In other words, everything that could alter the analysis (tokenization, white space, stop words, etc.) has been removed. To locate the two frames we have chosen to use the supervised algorithm SVM (Support Vector Machines). We decided to use SVM because it is one of the most widely used supervised text classification algorithms in industry and academia (Joachims, 1998). SVM is based on the idea that any linear model is valid to classification if the classes are linearly separable, suffice to find a hyperplane that discriminates both sets. In other words, any regression technique can be used for classification if we separate a sample in two groups: one group for trading and another where the value of the regression is calculated and assigned to the corresponding class (García-Marín and Calatrava, 2018).

The result of the application of the algorithm was the total coding of the sample taking as training the 200 articles coded by the authors (75% training 25% test). The result reached 80% reliability and 87% AUC, being really significant data. The kernel function used was linear ($c = 1.30$; $\varepsilon = 0.10$). All analyses have been performed under Orange Data Mining software (Demšar et al., 2013) under Python 3. As its developers point out Orange is, together with KNIME, one of the easiest to use data mining programs (Demšar & Zupan, 2013). The flexibility of the software allows you to apply different text automation techniques quickly and easily so that the learning curve can be classified as moderate.

4 Findings

As mentioned above, this research presupposes the existence of security and human rights frames when reporting on migration crises. Of course, this presumption is not coincidental since we have an extensive bibliography which supports that the use of these narrative frameworks are very present in issues related to migration processes. An example is how immigration has been included as a topic on the security agendas, especially since the 9/11 attacks (Calvillo-Cisneros, 2019). In addition, in the Spanish case, some authors argue that Spanish policy has been aimed at the prevention, containment and retention of immigration (López-Sala & Moreno-Amador, 2020). However, to reinforce this issue, a cloud of words has been made from the headlines during both migration crises (Charts 1 and 2). Although this research focuses on the content of the news, and how the headlines have been constructed can be a significant indication of the use of frames.

The configuration of the keywords of both crises and the number of news dedicated suggests that visibility is more influenced by media elements than statistics. This idea is reinforced if we take into account that the Spanish press tends to explain the religiosity of immigrants from North African countries, not being so in the case of people from Latin America or Eastern Europe (García-Marín, 2015).

If we take into account the main key words of the crises we can see clear differences. However, it should be noted that the configuration of the headlines could have partly suffered from the clickbait phenomenon. In other words, using shocking and emotive terms to attract potential readers. Nevertheless, the use and frequency of specific terms is a very useful exploratory indicator for finding frames. On the one hand, Chart 1 shows some particularly interesting information elements such as the weight of words such as "*Rabat*", "*Morocco*", "*Minors*", "*Border*" or "*Sahara*". While on the other hand, Chart 2 shows concepts more expected within an environment of migration crisis, "*Immigrants*", "*Lesbos*", "*Camp*", "*Decontrol*" or "*Dinghy*". This first analysis indicates that the crisis in Ceuta and Melilla has been strongly influenced by other dynamics external to the continuous migratory flows to Spain. An example is the importance of terms and names that are related to issues associated with foreign policy, such as the case of Polisario Front leader Brahim Ghali who was hospitalized in Spain. It is therefore to be expected that the security frame will be much more predominant in this crisis than of that of the Canary Islands.

As for the content of the news, Table 1crisis shows the use of frames by newspapers in both crises. As can be see, the data are very significant. On the one hand, in the case of the in Ceuta and Melilla, all the newspapers mostly used the security frame (with *El País* being the medium which obtained the lowest score of 70.9%). On the other hand, the media behavior on immigrants arriving on the islands of the Canary Archipelago has undergone more striking changes. In the cases of *ABC* and *El Mundo*, the security frame is more present (between 55 and 60%) while in *El País* and *La Vanguardia* the human rights frame is the predominant one (63 and 62%). Of course, it should be noted that since the frames are not mutually exclusive, we can establish that both are deeply rooted in the journalistic routines of the editorial

Chart 1 Word Cloud Ceuta and Melilla crisis (headlines). *Source* own elaboration

Chart 2 Word Cloud Canary Islands crisis (headlines). *Source* own elaboration

Table 1 Use of frames by newspapers

Ceuta and Melilla crisis					
Newspaper	N	Security		Human rights	
ABC	212	170	80.18%	113	53.30%
El Mundo	148	131	88.51%	67	45.27%
El País	220	156	70.90%	134	60.90%
La Vanguardia	104	85	81.73%	54	51.92%
Total	684	542	79.23%	368	53.80%
Canary Islands crisis					
ABC	53	31	55.35%	24	45.28%
El Mundo	48	29	60.41%	23	47.91%
El País	65	26	40.00%	41	63.07%
La Vanguardia	27	14	51.85%	17	62.96%
Total	193	100	51.81%	105	54.40%

Source Own elaboration

lines. However, these results coincide with the theses that argue that the traditional Spanish press usually identifies immigration as a problem (Nogales-Bocio, 2020).

As a result of the data, we can establish that the Spanish press has undergone a securitization process when reporting on the migratory phenomenon (H1). However, caution should be exercised with the extracted data. In our opinion, the coverage of the crisis in the Canary Islands presents data more in line with reality. Firstly, because the debate on immigration focuses more on the situation and the means of the security forces and on the capacity to welcome these immigrants. In addition, the situation of minors and the procedure for the deportation of illegal immigrants according to international law: "The Canary Islands should not be a new Lesbos, *La Vanguardia*, 11/20/20", "Migratory chaos, *El País*, 20/20 / 11 ", "Police, mayors and Government of the Canary Islands rebel against the management of Sánchez, *ABC*, 11/20/20". And secondly, the crisis in Ceuta and Melilla has been strongly framed on the responsibility of the Moroccan Government. In this respect, we believe that the high percentages of the security frame in this matter are more related to foreign policy issues than to the problem of immigration itself. This statement is based on the amount of news that raise the need for a deeper debate on the European Union's migration policy and the use of immigration as an element of pressure by Morocco: "Morocco pressures Sánchez with a migratory wave over Ceuta, *El Mundo*, 18/05/21", "Brussels remembers that Spanish borders are European borders, *La Vanguardia*, 19/05/21", "It is an invasion consented by Morocco, *ABC*, 20/05/21".

5 Discussion and Conclusion

In conclusion, verifiable differences are observed depending on the case study. There is a very homogeneous media behavior when it comes to reporting on migratory movements in the cities of Ceuta and Melilla. As can be seen, the security frame is very predominant in all the media analyzed. Therefore, our hypothesis (H1) can be considered positive but with nuances. In our opinion, the coverage of the crisis in the autonomous cities is strongly related to diplomatic tensions with the Moroccan government. In other words, given the media, migratory flows can cause serious security problems if Morocco does not actively participate in their prevention. In fact, in this case, we find more news that focuses on the role of the European Union and the need to coordinate migration policies, especially in the countries of southern Europe.

Another interesting fact is the difference in frames in the case of the crisis in the Canary Islands. The results show two distinct groups between conservative and progressive-leaning newspapers, which is not very surprising. However, the fact that such differences exist has been particularly remarkable from a theoretical point of view. On the one hand, when state actors (governments of North African countries) are directly involved in the migration phenomenon, the media tend to securitize the problem. Moreover, on the other hand, unlike armed conflicts, the media have reporters and newsrooms on the ground. This direct contact with people crossing the border allows us to create more empathetic news that gives us the voice to multiple stories, in many cases dramatic.

The results are consistent with the existing academic literature. In the case of the Canary Islands, it can be concluded that ideology is a variable of special importance when it comes to encoding news. The relevance of editorial lines has been evident in multiple investigations (Igartua et al., 2005; Santos, 2020). On the other hand, the tensions in Ceuta and Melilla have the added peculiarity of being part of a diplomatic crisis. In this sense, we can highlight that the role of Morocco presents in journalistic routines. An example is how the media in France and Spain act with some prudence when reporting on the Sahara conflict (Moreno-Mercado, 2020). Nevertheless, various news reports blame the Spanish government for fueling the crisis by allowing Ghali to be hospitalized on Spanish territory.

Finally, we want to highlight the potential of automatic word processing in studies of this type. We subscribe to the words of Serrano-Contreras (2021): "We must translate the unstructured information into a set of data frames to be structured; this is essential for producing quantitative data from qualitative information". The use of these techniques allows us to analyze large volumes of information in a short time. This research aims to make a small contribution to the subject of study without forgetting its limitations since immigration presents multiple visions within Spanish society (Fierro and Paella, 2021).

References

Abad-Quintanal, G. (2015). El concepto de seguridad: su transformación. *Comillas Journal of International Relations, 4*, 41–51.

Arango, J. (2007). Las migraciones internacionales en un mundo globalizado. *Vanguardia Dossier, 22*, 6–15.

Arango, J., Ramón, M., Moya-Malapeira, D., & Sánchez-Montijano, E. (2019). Inmigración, elecciones y comportamiento político. *Anuario CIDOB de la Inmigración,* 1–15.

Buzan, B., Waever, O., & Wilde, J. (1998). *Security. A new framework for analysis*. Lynne Rienner Publishers.

Calvillo-Cisneros, J. M. (2019). La situación actual de los refugiados a la luz del ordenamiento jurídico internacional. *Vergentis: Revista de Investigación de la Cátedra Internacional Conjunta Inocencio III,* (9), 219–249.

Demšar, J., Curk, T., Erjavec, A., Gorup, C., Hočevar, T., Milutinovič, M., Možina, M., Polajnar, M., Toplak, M., Starič, A., Štajdohar, M., Umek, L., Žagar, L., Žbontar, J., Žitnik, M., & Zupan, B. (2013). Orange: Data mining toolbox in Phyton. *The Journal of Machine Learning Research, 14*(1), 2349–2353.

Demšar, J., & Zupan, B. (2013). Orange: Data mining fruitful and fun-a historical perspective. *Informatica, 37*(1), 55–60.

Entman, R. M. (1993). Framing: Toward clarification of a fractured paradigm. *Journal of Communication, 43*(4), 51–58.

Fajardo-Fernández, R., & Soriano-Miras, R. M. (2016). La construcción mediática de la inmigración en el Mediterráneo: ¿no-ciudadanía en la prensa española? *Revista Internacional De Estudios Migratorios (RIEM), 6*(1), 141–169.

Fierro, J., & Parella, S. (2021). Social trust and support for immigrant's social rights in Spain. *Journal of Ethnic and Migration Studies*, 1–17.

García-Marín, J. (2015). La inmigración musulmana en la prensa española. In: C. De Cueto-Nogueras, & J. García-Marín (Eds.), *La mirada del otro: percepciones del islam* (pp. 165–185). Plaza y Valdés.

García-Marín, J., & Calatrava-García, A. (2018). The use of supervised learning algorithms in political communication and media studies. Locating Frames in the Press. *Communication & Society, 31*(3), 175–188.

Huysmans & Squire, J. V. (2009). Migration and Security. In: En M. a. Dunn Cavelty, *Handbook of Security Studies* (pp. 169–179). Routledge.

Igartua, J. J., Muñiz, C., & Cheng, L. (2005). La inmigración en la prensa española. Aportaciones empíricas y metodológicas desde la teoría del encuadre noticioso. *Migraciones, 17*, 143–181.

Joachims, T. (1998). *Making Large-Scale SVM Learning Practical*. Technical Report.

López-Sala, A., & Moreno-Amador, G. (2020). En busca de la protección a las puertas de Europa: refugiados, etiquetado y prácticas disuasorias en la frontera sur española. *Estudios Fronterizos,* (21), 1–20

Oller-Alonso, M., Blanco-Herrero, D., Splendore, S., & Arcila-Calderón, C. (2021). Migración y medios de comunicación. Perspectiva de los periodistas especializados en España. *Estudios sobre el Mensaje Periodístico, 27*(1), 205–228.

Moreno-Mercado, J. M. (2020). La seguridad como encuadre genérico de los conflictos: El conflicto del Sáhara Occidental en la prensa de España y Francia (2014–2019). *Revista De Estudios Mediterráneos (REIM), 29*, 114–129.

Moreno-Mercado, J. M., García-Marín, J., & Luengo, O. G. (2021). El conflicto de Siria en la prensa española: un análisis sobre la securitización de la guerra. *Política y Sociedad, 58*(3).

Nogales-Bocio, A. I. (2020). Periodismo low, periodismo slow y derechos humanos. Diferencias y riesgos de la cobertura informativa del fenómeno migratorio en España. *Revista Inclusiones, 7*(2), 75–103.

Nye, J., & Keohane R. (1989). *Power and interdependence: World politics in transition*. Little, Brown and Company.

Santos, T. (2020). *Framing of refugees in Spanish newspapers*. Lunds Universitet.
Serrano-Contreras, I. (2021). NPL on YouTube: A look a Feminism. In M. Musial-Karg, & O. G. Luengo (Eds.) *Digitalization of democratic processes in Europe* (pp. 127–138). Springer
Waltz, K. (1979). *Theory of international politics*. McGraw Hill.
Wendt, A. (1999). *Social theory of international politics*. Cambridge University Press.

José-Manuel Moreno-Mercado is PhD Candidate in Department of Political Science at the University of Granada. Graduate in Political Science by University of Granada and Master's degree in Peace Culture, Conflicts, Education and Human Rights by Institute of Peace and Conflicts (IPAZ) and the same university. He won the I Juan del Pino Award for the best TFM (Master dissertation) in Applied Sociology with the title: "Locating frames with Support Vector Machines (SVM): The behavior of the media in the conflicts of Yemen and Ukraine" awarded by the University of Malaga. Likewise, He has been awarded a scholarship from the Fundación Seminario de Investigación para la Paz (SIP) in Zaragoza. Also is author of several articles and collaborations on books about Media and Conflict Studies and has participated in numerous conferences.

Adolfo Calatrava-García is a professor in the Department of International Relations and Global History of the Complutense University of Madrid. PhD in Political Science from the University of Granada. Diploma of Advanced Studies (DEA) in Economics and International Relations from the Autonomous University of Madrid. Master in Security and Defence from the Complutense University of Madrid and CESEDEN. His research focuses on studies of international security, geopolitics, political transitions in the ex-Soviet space and in the Mediterranean and middle east. He has been a professor at the Carlos III University, the Nebrija University and the University of Granada. In addition, he has worked as an advisor in the Secretary of State for Security of the Ministry of the Interior, training advisor and project technician in the Euro-Arab Institute of Education and Training (INSTEA), and project technician in the Euro-Arab Foundation for Higher Studies. He has been a visiting researcher at Cardiff University, the London School of Economics and the University of Tromsø. In 2007, he was a fellow of the American German Marshall Foundation in the Memorial Marshall Fellowship program for young European leaders.

José-Miguel Calvillo-Cisneros holds a PhD in International Relations (2010) from the Complutense University of Madrid (UCM) and a Degree in Political Science and Administration from the same university with a specialty in International Studies. He studied the Master's Degree in International Solidarity Action of Europe at the Carlos III University of Madrid with the titles of expert in Development Cooperation; humanitarian action; and Immigration, refuge and asylum. They have numerous scientific publications and research in their main lines of research: the link between security and development; international cooperation, humanitarian action, international migration and international conflicts, especially in Afghanistan. Apart from academic life, he has more than 15 years of experience working in the field of international cooperation for development that has allowed him to know the field of numerous countries such as Afghanistan, Tajikistan, Egypt, Tanzania, Nicaragua, El Salvador, among others.

Index

J. Cayón Peña (ed.), *Security and Defence: Ethical and Legal Challenges in the Face of Current Conflicts*, Advanced Sciences and Technologies for Security Applications,
https://doi.org/10.1007/978-3-030-95939-5

The manufacturer's authorised representative in the EU is Springer Nature Customer Service Centre GmbH, Europaplatz 3, 69115 Heidelberg, Germany. If you have any concerns regarding our products, please contact ProductSafety@springernature.com

Printed and bound by CPI Group (UK) Ltd, Croydon, CR0 4YY
30/09/2026
02225496-0004